PRODUCING PRECARITY

POSTMILLENNIAL POP SERIES

General Editors: Karen Tongson and Henry Jenkins

Puro Arte: Filipinos on the Stages of Empire
Lucy Mae San Pablo Burns

Media Franchising: Creative License and Collaboration in the Culture Industries
Derek Johnson

Your Ad Here: The Cool Sell of Guerrilla Marketing
Michael Serazio

Looking for Leroy: Illegible Black Masculinities
Mark Anthony Neal

From Bombay to Bollywood: The Making of a Global Media Industry
Aswin Punathambekar

A Race So Different: Performance and Law in Asian America
Joshua Takano Chambers-Letson

Surveillance Cinema
Catherine Zimmer

Modernity's Ear: Listening to Race and Gender in World Music
Roshanak Kheshti

The New Mutants: Superheroes and the Radical Imagination of American Comics
Ramzi Fawaz

Restricted Access: Media, Disability, and the Politics of Participation
Elizabeth Ellcessor

The Sonic Color Line: Race and the Cultural Politics of Listening
Jennifer Lynn Stoever

Diversión: Play and Popular Culture in Cuban America
Albert Sergio Laguna

Open TV: Innovation beyond Hollywood and the Rise of Web Television
Aymar Jean Christian

Antisocial Media: Anxious Labor in the Digital Economy
Greg Goldberg

More Than Meets the Eye: Special Effects and the Fantastic Transmedia Franchise
Bob Rehak

Spreadable Media: Creating Value and Meaning in a Networked Culture
Henry Jenkins, Sam Ford, and Joshua Green

Playing to the Crowd: Musicians, Audiences, and the Intimate Work of Connection
Nancy K. Baym

Old Futures: Speculative Fiction and Queer Possibility
Alexis Lothian

Anti-Fandom: Dislike and Hate in the Digital Age
Edited by Melissa A. Click

Social Media Entertainment: The New Intersection of Hollywood and Silicon Valley
Stuart Cunningham and David Craig

Video Games Have Always Been Queer
Bo Ruberg

The Power of Sports: Media and Spectacle in American Culture
Michael Serazio

The Race Card: From Gaming Technologies to Model Minorities
Tara Fickle

Open World Empire: Race, Erotics, and the Global Rise of Video Games
Christopher B. Patterson

The Content of Our Caricature: African American Comic Art and Political Belonging
Rebecca Wanzo

Stories of the Self: Life Writing after the Book
Anna Poletti

The Dark Fantastic: Race and the Imagination from Harry Potter to the Hunger Games
Ebony Elizabeth Thomas

Hip Hop Heresies: Queer Aesthetics in New York City
Shanté Paradigm Smalls

Digital Masquerade: Feminist Rights and Queer Media in China
Jia Tan

The Revolution Will Be Hilarious: Comedy for Social Change and Civic Power
Caty Borum

The Privilege of Play: A History of Hobby Games, Race, and Geek Culture
Aaron Trammell

Unbelonging: Inauthentic Sounds in Mexican and Latinx Aesthetics
Iván A. Ramos

Sonic Sovereignty: Hip Hop, Indigeneity, and Shifting Popular Music Mainstreams
Liz Przybylski

Style: A Queer Cosmology
Taylor Black

Normporn: Queer Viewers and the TV That Soothes Us
Karen Tongson

Where the Wild Things Were: Boyhood and Permissive Parenting in Postwar America
Henry Jenkins

Fandom for Us, by Us: The Pleasures and Practices of Black Audiences
Alfred L. Martin, Jr.

Producing Precarity: The Costs of Making TV in Poor Places
Curtis Marez

Producing Precarity

The Costs of Making TV in Poor Places

Curtis Marez

NEW YORK UNIVERSITY PRESS
New York

NEW YORK UNIVERSITY PRESS
New York
www.nyupress.org

© 2025 by New York University
All rights reserved

Please contact the Library of Congress for Cataloging-in-Publication data.

ISBN: 9781479836703 (hardback)
ISBN: 9781479836727 (paperback)
ISBN: 9781479836741 (library ebook)
ISBN: 9781479836734 (consumer ebook)

This book is printed on acid-free paper, and its binding materials are chosen for strength and durability. We strive to use environmentally responsible suppliers and materials to the greatest extent possible in publishing our books.

The manufacturer's authorized representative in the EU for product safety is Mare Nostrum Group B.V., Mauritskade 21D, 1091 GC Amsterdam, The Netherlands. Email: gpsr@mare-nostrum.co.uk.

Manufactured in the United States of America

10 9 8 7 6 5 4 3 2 1

Also available as an ebook

Dedication

For Shelley

CONTENTS

List of Figures xi

1. Cold Opening 1
2. Poor Locations 30
3. Crime TV 53
4. Horror TV 72
5. Science Fiction TV 95
6. Black Atlanta TV 118
7. Gentrification TV 144

   Afterword: Streaming Poor Places 173

   Acknowledgments 181

   Television Shows Cited 183

   Notes 187

   Bibliography 211

   Index 229

   About the Author 235

LIST OF FIGURES

CHAPTER 1

Figure 1.1: Amara on the set of a movie called *The Admiral's Mistress*, on the TV show *Get Shorty*. 1

Figure 1.2: Drug Queen Amara de Escalones (Lidia Porto), reading the script for *The Admiral's Mistress* (*Get Shorty*). 2

Figure 1.3: Marisol Suarez (Ana Ortiz) prepares to clean the murder scene of another Latinx maid on *Devious Maids*. 7

Figure 1.4: A prison fight on *Will Trent*, shot at Gwinnett County Detention Center. 12

Figure 1.5: TV shows within TV shows: *The Curse*. 17

CHAPTER 2

Figure 2.1: Lucifer and a drug lord named El Jefe plan a murder inside an Albuquerque Rail Yards building in *The Messengers*. 32

Figure 2.2: The Albuquerque Rail Yards. 33

Figure 2.3: Shadowbox Studios, bordered by waterways, forests, and the Georgia Department of Corrections or "Cop City," in an area over 70% Black. 36

Figure 2.4: "Stop the Swap" poster opposing the land exchange between the city of Atlanta and Shadowbox/Black Hall Studios, www.stoptheswap.org. 39

Figure 2.5: Thony prepares to clean a warehouse murder scene in *The Cleaning Lady*. 45

Figure 2.6: DEA agents and white supremacist meth dealers face off at the Navajo settlement To'hajiilee in *Breaking Bad*. 50

CHAPTER 3

Figure 3.1: Jonathan Banks (back, right), who plays Mike Ehrmantraut on *Breaking Bad* and *Better Call Saul*, with members of the Albuquerque Police Department. 53

Figure 3.2: The watchtower of Santa Fe, New Mexico's Old Main Penitentiary in *MacGruber*. 58

Figure 3.3: The Old Main Penitentiary in *The Lost Room*. 58

Figure 3.4: Albuquerque Municipal Detention Center in *Breaking Bad*. 59

Figure 3.5: Kim Wexler leaves the Penitentiary of New Mexico in Santa Fe after visiting Saul Goodman in *Better Call Saul*. Such high angle shots of prisoners and penitentiary yards mimic prison surveillance. 60

Figure 3.6: A prisoner on his way to the visitation room in the Gwinnett County Detention Center, Lawrenceville, Georgia, on the show *Ozark*. 60

Figure 3.7: Trans Deputy Brianna Bishop fists bumps a fellow officer on *Deputy*. 64

Figure 3.8: On *The Cleaning Lady*, Thony is placed in ICE detention. 70

Figure 3.9: ICE-detained children watching TV, *The Cleaning Lady*. 71

CHAPTER 4

Figure 4.1: Serial killer Dre cleans up a murder scene in the home of her upper-class victim on *Swarm*. 81

Figure 4.2: Having killed one white supremacist, the vampire Lemuel (a former slave) prepares to dispatch a second on *Midnight, Texas*. 83

Figure 4.3: A Latina maid pushes a cleaning cart at Albuquerque's Sundowner Motel in *Preacher*. 86

Figure 4.4: Tic and Leti in the forest as a racist policeman approaches in *Lovecraft Country*. 90

CHAPTER 5

Figure 5.1: Sister Night from *Watchmen*. 98

Figure 5.2: Superman hovering over a pedestal like a Confederate memorial in *Naomi*. 107

Figure 5.3: Thunder destroys a Confederate monument, *Black Lightning*. 108

Figure 5.4: Sherriff Joy Hawk, *Outer Range*.   116
Figure 5.5: The buffalo return, *Outer Range*.   117

CHAPTER 6
Figure 6.1: Bishop Greenleaf welcomes the police, on *Greenleaf*.   119
Figure 6.2: A policeman shot on *Greenleaf*.   120
Figure 6.3: Grace Greenleaf looks on as the policeman dies.   120
Figure 6.4: Mavis McCready behind the bar in *Greenleaf*.   143

CHAPTER 7
Figure 7.1: Netflix's Mesa del Sol housing development, with soundstage 3 visible in the distance.   147
Figure 7.2: Trilith Studios and housing development.   148
Figure 7.3: Alexandria, *The Walking Dead* housing development.   150
Figure 7.4: "Wash Her Out!" Anti-gentrification activist Yoli bombs Lynn Hernandez with laundry detergent.   162
Figure 7.5: A settler colonial memorial on Universal's Court House Square in *Rutherford Falls*.   169

1

Cold Opening

Get Shorty (Epix, 2017–19), a TV show inspired by the Elmore Leonard novel of the same name, presents a meta-reflection on the process of media-making in poor places. Season 1 deals with Amara de Escalones (Lidia Porto), a ruthless Guatemalan casino boss and drug trafficker in Puyallup, Nevada, who launders money by investing in a movie. (See Figure 1.1.) The episode titled "Shooting on Location" depicts the production of *The Admiral's Mistress*, an English period romance set during the Napoleonic wars.

Amara funds the film to launder drug money, and she convinces the studio to shoot it in Nevada by providing armed Latinx gangsters to do free labor on set. With a penchant for animal prints and sharp red nails, Amara beds and then murders the owner of a water park where she launders money before forcing herself on the terrified producer of *The Admiral's Mistress*. (See Figure 1.2.) While *Get Shorty* also depicts a

Figure 1.1: Amara on the set of a movie called *The Admiral's Mistress*, on the TV show *Get Shorty*.

1

Figure 1.2: Drug Queen Amara de Escalones (Lidia Porto), reading the script for *The Admiral's Mistress* (*Get Shorty*).

Hollywood studio head who defrauds insurance companies, Amara represents a kind of racialized and gendered terror that makes Hollywood's transgressions pale by comparison.

Although set in Nevada, "Shooting on Location" was filmed on a backlot in Albuquerque to take advantage of state subsidies. The film tax incentive program and other subsidies in New Mexico—a poor state of mostly Mexican and Native American people—effectively diverts taxes from social welfare to media industries in Los Angeles. To return to *Get Shorty*, via state subsidies TV producers redistribute wealth upwards in ways that seem unethical if not criminal, but the racist and sexist representation of Amara distracts from the local production context and Hollywood's complicity in the theft of public resources. *Get Shorty* exemplifies my claim that the content and production process of TV shows made in poor parts of the United States constitute powerful, interrelated forms of racial capitalism in place.

If racial capitalism names capitalism's development in racial directions, "racial capitalism in place" describes those developments locally, in places where media are made. In this book I elaborate a theory of TV racial capitalism in place by turning to shows shot in Georgia, New Mexico, and a few in East Los Angeles, to take advantage of local tax incentives and other state subsidies. They are relatively poor states and regions with large Indigenous, Black, and Latinx populations, exemplifying extreme forms of historic and ongoing structural inequality. With origins in colonial and plantation histories, racial capitalism in these different places is now partly centered in film and TV production.

A total of 60 countries offers financial incentives as part of a global competition to attract Hollywood productions, with tax rebates between 20% and 40%. This includes 32 countries in Europe (both relatively wealthy ones—France, Germany, the UK, and relatively poor ones—Serbia, Hungary, Ireland); 12 in Asia/Oceania (such as New Zealand, India, Malaysia, South Korea, and Taiwan); 7 in Africa and the Middle East (South Africa, Israel, and Saudi Arabia, for example); and 7 in Latin America and the Caribbean (such as Colombia, the Dominican Republic, and Trinidad & Tobago). Meanwhile, all 10 Canadian provinces and 36 US states provide subsidies for film and TV production. Encouraged by the competitive international subsidy systems, Myles McNutt argues that film and TV shows are increasingly "mobile productions."[1]

The global landscape of state support is complicated and heterogeneous, with multiple determinations and local, historical, and material contexts. Here I focus on a few pieces of that larger puzzle, starting with TV production in the United States. As Vicki Mayer reminds us in her book about state filming subsidies in Louisiana, *Almost Hollywood, Nearly New Orleans: The Lure of the Local Film Economy*, states use public subsidies to "compete in a race to the bottom for a film economy."[2] Overrepresented on the list of film subsidy providers are deindustrialized, rust belt states (Illinois, Indiana, Kentucky, Missouri, Ohio, Pennsylvania, New York, and West Virginia), and Sunbelt states with significant Black, Latinx, and Indigenous populations (Arizona, Arkansas, California, Colorado, Georgia, Louisiana, Mississippi, New Mexico, North Carolina, South Carolina, and Texas). Many of the states are poor, particularly Louisiana, Mississippi, and New Mexico, the three states with the highest percentage of residents living below the poverty line.[3] Out of all the possible locations for a study of subsidies, I've chosen to focus mostly on Georgia and New Mexico because they have been especially successful—to the detriment of their people—at leveraging local conditions of racialized poverty to appeal to Hollywood.

The first New Mexico film commission was founded in 1968, subsequently becoming the New Mexico Film Office, but film incentive programs began under Republican Governor Gary Johnson (R), who in 2002 signed a bill offering a 15% film production tax credit. His successor, Governor Bill Richardson (D), raised the tax credit to 25%.[4] In 2011, Governor Susana Martinez signed what was popularly referred to as the "Breaking Bad Bill," raising the tax rebate to 30%.[5] In 2019, Governor Michelle Lujan Grisham (D) signed a new bill that gave incentives to companies to enter public/private relationships to purchase or lease land and build studio infrastructures. The bill raised the cap on what the state can pay out from $50 million to $110 million. It also offered a rural "uplift credit" of 5% for productions at least 60 miles outside the Bernalillo and Santa Fe County corridor. New Mexico currently offers a combination of tax rebates that can total as high 40%.[6] New Mexico's incentive program paid out more than $110 million in 2023, with a projected payout of $160 million by 2028.[7]

While serving as governor, Jimmy Carter established Georgia's first film commission in 1973, and in the 1980s, Georgia aggressively attempted to attract film and TV production by advertising lower costs for locations, labor, and housing than other places.[8] In 2005, Georgia instituted its first film incentive, a 9% tax rebate that was effectively increased to 30% in 2008, including costs for actors, writers, directors and showrunners—in other words, the most privileged media workers.[9] During the COVID-19 pandemic, Georgia's incentives totaled $870 million, even while the state cut funding for public education by almost $1 billion.[10] Film and TV production expanded after the pandemic, with Georgia paying out $1.3 billion in filming subsidies in 2023.[11]

Georgia and New Mexico further attract media makers by providing public lands at significantly reduced rates for the building of soundstages and related facilities, and by subsidizing infrastructure such as roads, sewers, and the electric power grid.[12] The state of California and the City of Los Angeles are also in the game, providing generous subsidies to TV producers filming in the historically poor but recently gentrifying Latinx neighborhood of Boyle Heights. In addition to direct subsidies for production, states and municipalities enter public-private partnerships that effectively subsidize the construction of studios, roads, and mixed-use residential developments. "Cultural industries use the aura of their operations and products," Mayer writes, as well as unrealistic promises of jobs, "as leverage to reduce their economic costs" in poor places, resulting in "the extreme concentration of wealth under the twin banners of economic and cultural renewal."[13]

Hollywood and local state officials promise more jobs and more spending on local businesses, but such benefits are minimal compared to profits. In his national study of different states incentive programs, economist John Charles Bradbury found

> no link between incentives that promote in-state filming and states' economic growth or level. The findings are not surprising given previous estimates of small and nonrobust effects. . . . The empirical findings are also consistent with reports by state economic development agencies that have estimated weak financial returns to their film incentive programs. . . . In total, the findings do not support the economic development justification often used to promote state subsidies to the film industry. . . . If resources are going to film production companies but are not having a positive effect on local economies, then the benefits appear to be flowing entirely outside the state.[14]

This conclusion was confirmed by Bradbury's study of incentives in Georgia. The flow of benefits outside the state was acknowledged by Ryan Millsap, president of Blackhall Studios (recently re-branded as Shadowbox Studios) in the Atlanta area, who explained to a local reporter, "We're taking movies that were conceived in LA, funded in LA, and then shipped to Georgia for manufacturing. Then they're brought back to LA for distribution (and) then all the money stays in Los Angeles after the movie's made."[15]

Patrick Button's study of incentives in Louisiana and New Mexico also debunks optimistic claims about jobs and economic growth. He argues that data generated by TV industry stakeholders is often unrealistic and inflated. It is in Hollywood's interest to cite impressive job numbers to justify state subsidies. But Button argues that their actual economic impact is minor. The increase in employment is not "statistically significant" and "very costly per job created." Media makers and state officials also claim that subsidies stimulate other segments of the local economy, especially tourism. Button concedes the point but emphasizes that, because the direct economic impact of location shooting is so small, any "spillover effects on other industries" are also small. I would add that tourism in New Mexico means low-wage, racialized, and gendered jobs for Mexican and Indigenous people in hotels and restaurants, a situation that is consistent with the region's longer history as a site of Spanish and then Anglo settlement and colonial labor relations.

Indeed, in New Mexico and Georgia, state subsidies have driven the growth of cleaning businesses that cater to film and TV producers by providing mostly Black, Latinx, and Indigenous workers to clean production offices, soundstages, and filming locations. Atlanta's Clean Tu Casa, for example, promises that its workers "discreetly" clean studios, production offices, warehouses, kitchens and break rooms, and restrooms, while Georgia Clean Trauma Services provides scene-change cleanup and post-production cleanup, with a specialty in cleaning staged crime scenes (they also clean real crime scenes).[16] Meanwhile, Albuquerque's Blue Moon Cleaning and Restoration promotes its workers' "superior set cleaning, basecamp cleaning and damage restoration," including cleaning up after staged scenes of violence: "Whether dealing with crash and explosion scenes, shootout scenes, or close-quarter bloody fight scenes, we've seen it all."[17] Similarly, the workers at Albuquerque's Green Sweep provide a number of services for film and TV, including pre and post-production office cleaning, house cleaning for talent and crew, and "specialty cleans including warehouses, fake blood clean-up, and cleaning on sites without water or power."[18] Other cleaning companies serving film and TV production include Atlanta's Millennium Facility Film and Television Studio Cleaning Service, and Santa Fe's Sun Green Cleaner.[19] Most of these cleaning companies' websites feature images of their workers, who largely appear to be Black, Latinx, and Indigenous women, suggesting the kinds of low-wage, gendered, and racialized labor that Hollywood brings to poor places. The Atlanta- and Albuquerque-made shows *Devious Maids* and *The Cleaning Lady* fictionalize cleaning workers with characters forced to clean crime scenes. Indeed, both shows incorporate scenes of sponges and mops wiping away blood in their title sequences and other scenes. (See Figure 1.3.)

TV productions in Georgia and New Mexico and other poor places indirectly benefit from low labor costs in service and agricultural sectors, but employ relatively few local workers themselves. *The Los Angeles*

Figure 1.3: Marisol Suarez (Ana Ortiz) prepares to clean the murder scene of another Latinx maid on *Devious Maids*.

Times reports that, over its 62 episodes, *Breaking Bad*, one of the most popular and profitable shows shot in New Mexico, employed only 88 local workers. Many of the relatively small number of local positions created are poor. As Mayer writes, the jobs that boosters use to "justify starving other state spending priorities" are "temporary. Many film jobs last only a day. They are unstable. Film jobs flourish where they receive the most subsidy, often in right-to-work states that erode the power of the industry's own labor unions and professional guilds. Few people make their sole living in the role of a film and television worker."[20] (Georgia is a right-to-work state but New Mexico isn't). Button concludes by suggesting that New Mexico state funds could have been better spent on "public goods such as education, health, or infrastructure."[21] While subsidy boosters present a rosy job picture, TV shows sometimes hint at the truth. The Georgia-made *Being Mary Jane* uses one character's job guarding film trailers as a sign of his tragic decline (chapter 6). Similarly, when a former high-cost escort is relocated to Albuquerque as part of witness protection in *The Girlfriend Experience* (2016–2021), her sad job on the bottling line at a local brewery (Albuquerque's Marble Brewery) drives her to drink and ultimately return to sex work (S2E2). These brief scenes give us revealing glimpses of TV racial capitalism in place.

TV Racial Capitalism in Place

Coined by South African scholars, the concept of racial capitalism was most influentially elaborated by Cedric Robinson in *Black Marxism: The Making of the Black Radical Tradition* (1983), where he argued that capitalism is inextricably fused with racism. "Racial capitalism," Robinson writes, means that "as a material force," racism "permeate[s] the social structures emergent from capitalism." As such, both "the development, organization, and expansion of capitalist society" and its corresponding "social ideology" have historically "pursued essentially racial directions."[22] Racial capitalism, in other words, combines systems of racialized theft and labor exploitation with the production and dissemination of the racist representations that support them. Robinson further elaborated his theories of racial capitalism in *Forgeries of Memory and Meaning: Blacks and the Regimes of Race in American Theater and Film Before World War II* (2007). While Robinson's ideas have been influential in Black studies, Indigenous studies, ethnic studies, and American studies, where scholars have elaborated theories of racial capitalism as a settler colonial formation with different incarnations across time and space, such work has not taken up *Forgeries*.[23] His work on film has also been underappreciated in media studies.[24] *Producing Precarity: The Costs of Making TV in Poor Places* is thus the first book-length study to build on Robinson's history and theory of racial capitalism in US film.

In *Forgeries of Memory and Meaning*, Robinson provides a compelling account of the origins and development of the Hollywood film industry as a powerful form of racial capitalism. Robinson historicizes US cinema as responding to several challenges faced by an older racial regime, including the end of reconstruction, cross-racial labor organizing, and capitalist exploitation of European immigrants incompletely schooled in US racism. The challenges were also global, including the Mexican revolution, which threatened US finance capital. "Spurred by the powerful interests implicated in the formulation of the new racial regime" that emerged in the late nineteenth and early twentieth centuries, movie companies helped forge the "cultural discipline" and "social habituation" of a new Jim Crow racial regime. The president of American Mutoscope, precursor to the famous Biograph Company where D. W. Griffith worked, was a prominent railroad executive; its vice president was a national bank examiner; and its board included a coal company executive. The railroads and allied industries had insatiable appetites for Indigenous land and for disposable, racialized labor, and filmmakers legitimated such theft and exploitation with racist representations. The large body of early anti-Black, anti-miscegenation movies was part of broader capitalist investments in segregation as labor control, while movies about Mexican criminality and gender/sex perversity formed the dominant US visual culture of counterinsurgency.[25] Robinson concludes that racist movies justify the exploitation of workers of color and encourage racism among white workers. Racism historically precluded interracial solidarity and promoted forms of white supremacy as a psychic compensation for class differences among white people. Pictures vilifying people of color further normalize the disciplining of Black and Latinx labor and encourage white racism to the benefit of finance capitalists invested in film.[26]

In his discussion of jungle movies from the late 1920s and 1930s, for example, Robinson suggests that Hollywood is an industry of racial capitalism that intermixes racial inequalities in both representation and production. Jungle pictures helped legitimate the exploitation of workers in Asia, Africa, Latin America, and the Caribbean, but also, he implies, in Black Los Angeles. Robinson cites a 1929 article published in *Opportunity: A Journal of Negro Life* describing the thousands of Black extras employed to play Africans, Asians, and Pacific Island "savages." They were recruited from South Central Los Angeles to work across town in Culver City, where the all-white studios were located.[27] Like other municipalities in greater Los Angeles, racially restrictive covenants were common there. Culver City was a notorious sundown town and local police had a reputation for racist harassment. It was also home to many members of the Ku Klux Klan.[28] At the same time as they faced possible racist violence, Black extras were often required to perform extreme physical feats in unsafe conditions and to participate in racist representations—conditions that inspired sit-down strikes and other labor actions.[29] Black extras played disposable African diamond miners (*Diamond Handcuffs*, 1928) and ivory porters (*West of Zanzibar*, 1928). Although attention to place is implicit in Robinson's materialist perspective, his discussion of jungle movies is the rare example where he explicitly considers Hollywood's local context of racialized labor exploitation. In this way, he connects film content rationalizing the exploitation of African workers to a local mode of production partly based in the exploitation of African American workers.

Similar claims can be made regarding the relationship between movies featuring Mexicans and the exploitation of Mexican workers in Southern California. During the first three decades of its twentieth-century history, Los Angeles depended on Mexican labor to build and maintain rail lines; construct the homes driving regional real estate speculation; harvest and pack citrus and other crops; and tend the region's lawns and gardens. Mexicans were also subject to segregation (including restrictive covenants) and police and vigilante violence. Hollywood employed a small number as extras, but only the whitest Mexican actors received credited roles. In response to the perceived threat of Mexican revolutionaries in Los Angeles, and with the advent of sound technologies, filmmaking became an indoor business by the 1930s as studios were increasingly walled off from surrounding Mexican social spaces and employed their own private police forces. The many silent and early sound movies featuring Mexican bandits symbolically consolidated anti-Mexican whiteness and disciplined the Mexican workers on whom Hollywood still depends.[30] To extrapolate from *Forgeries of Memory and Meaning*, then, Hollywood not only made movies that celebrated a Jim Crow/Juan Crow racial regime on behalf of other industries. Locally, old Hollywood was itself an industry of racial capitalism, and this history continues to shape the present with the export of Hollywood racial capitalism to places such as Georgia and New Mexico and localization there as the appropriation of public resources, the occupation of land, and the exploitation of racialized and gendered labor in the production of TV shows.

TV incentives, in other words, are part of what Jodi Melamed calls the "state-finance-racial violence nexus." For Melamed, racial capitalism presupposes collaborations between finance and state power to promote and protect capital accumulation. As she writes, "'state-finance-racial violence nexus' names the inseparable confluence of political/economic governance with racial violence, which enables ongoing accumulation through dispossession by calling forth the specter of race (as threat) to legitimate state counterviolence in the interest of financial asset owning classes." State racial violence in support of capitalism includes not only the police murder of Black and Latinx people but also the "letting die of the racialized poor."[31] From this perspective, by enabling accumulation at the expense of poor people of color, state programs that divert tax money from social welfare to Hollywood constitute forms of state violence.

TV production as racial capitalism in place also includes state violence in the form of local prisons and policing, a topic threaded through most of these chapters. Hollywood and local criminal justice systems are practical allies, forming a police and prison televisual complex.[32] New studios in New Mexico and Georgia are adjacent to police facilities and prisons, with shared interests in access to public land and infrastructure, as well as shared interests in the occupation of Indigenous lands. An infamous example is the symbiotic relationship between Blackhall/Shadowbox Studios and Atlanta's so-called Cop City police training facility in the occupation of Indigenous land, but the Netflix studio and the offices of its DEA and Homeland Security neighbors in Albuquerque also sit on Indigenous land (chapter 3). Inmates in the real prisons where TV shows film in Georgia and New Mexico are disproportionately Black and Latinx, and they disproportionately die there. In addition to using prisons as cheap, state-subsidized sets, TV producers hire off-duty police and prison guards as advisors, actors, and private security; and rent police vehicles and equipment. (See Figure 1.4.) Film studios, many with adjoining, mixed use residential communities, also rely on local police to protect their property in areas with large numbers of poor Black and Latinx people. For their part, prisons and the police benefit from their working relationships with film and TV producers. Hollywood makes payments to police departments and prisons for access to locations, equipment, and personnel, which they reinvest in new equipment or public-relations-enhancing charitable giving. Meanwhile a handful of police and prison staff create lucrative careers as celebrity advisors and actors. As Page and Ouellette suggest, prisons and police departments have been "reinvented" as part of an entrepreneurial culture industry. In these contexts, TV producers by and large represent the police (and implicitly, TV) as institutions that embody diversity.

One model for my study is research on the filming of HBO's *Treme* in New Orleans. In his essay on "the role of scripted cable television in the making and remaking of place in the conjuncture of post disaster

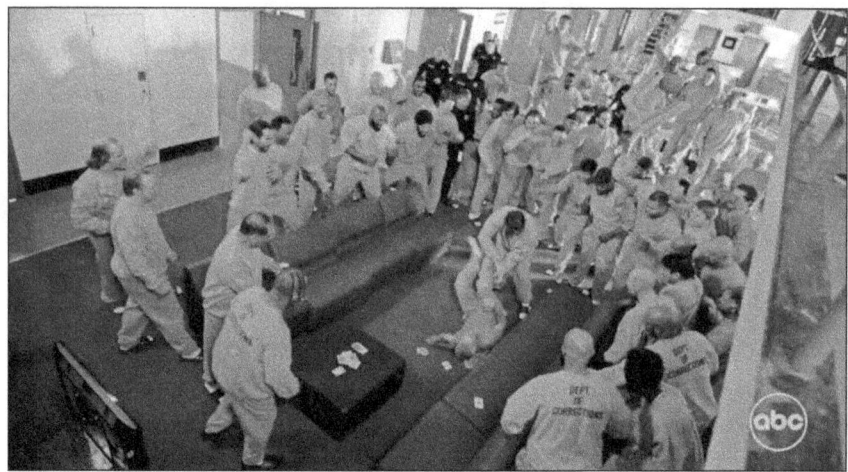

Figure 1.4: A prison fight on *Will Trent*, shot at Gwinnett County Detention Center.

crisis and the neoliberal transformation of urban space," Herman Gray claims that *Treme* helped remake post-Katrina New Orleans with representations of local authenticity (food, music and diversity) and narratives of individual enterprise that preclude a "critical engagement with public policy choices and state-centered redress for economic, cultural, and social injustice and inequality."[33] Helen Morgan Parmett similarly analyzes the show "as a site-specific spatial practice that plays a material role in rebuilding New Orleans" post Katrina along neoliberal lines, promoting private enterprise rather than state spending. Instead of focusing primarily on its content, Parmett traces "*Treme*'s spatial practices of production as they are implicated in on-location shooting, local hiring, charity, and tourism." She concludes that the program abets the abdication of "governmental responsibility for the care of its citizens, and for the maintenance and building of crucial infrastructure, as such labors are offloaded onto the private sector and citizens themselves and to the charitable contributions of the television industry and its viewers."[34] In *Almost Hollywood, Nearly New Orleans*, Vicki Mayer reaches a similar conclusion. While foregrounding the show's careful attention to its local context, its employment of locals, its appeal to local residents, and all the ways the people who made it gave back to the community in the form of charitable works, Mayer concludes that "*Treme* was no different from other entities that directed the excess of emotions after Katrina toward a philanthrocapitalism based on corporate efforts, private volunteerism, and cheap or free labor. . . . Popular media, including *Treme*, were central to this ideological mission by making recovery into a personal duty."[35] Much the same can be said of TV production in Georgia and New Mexico, where media makers combine upward wealth redistribution with individual acts of charity, poor substitutes for state social spending. Charity work, combined with "diverse" content, deflect from or justify Hollywood's appropriation of public resources.

In addition to studies of TV focused on race and political economy, I draw on research in critical geography, which analyzes how capitalism, state power, and cultural production make and transform different places.[36] In *A People's Guide to Los Angeles*, Laura Pulido, Laura Barraclough, and Wendy Cheng argue that rather than being fixed and natural, landscapes are dynamic and changing, the product of "millions of individual decisions, all made within the constraints of state policies and capitalist imperatives that are occasionally, and sometimes successfully, resisted by people, with an alternative vision of how the world should work."[37] Corporate and state institutions maintain the upper hand in part because their landscape-creating decisions are often obscured or deliberately hidden. As Pulido, Barraclough, and Cheng put it, "This is, in fact, one of landscape's greatest tricks and one of the most important ways in which landscape operates in the service of maintaining an unequal status quo. Because it is not always apparent why a landscape looks the way it does, it becomes easy to assume that it somehow naturally reflects the character, qualities, and moralities of the people who inhabit it."[38] If we substitute "filming location" for "landscape," we can see how TV locations "provide evidence about past generations, economic and political regimes, and ecologies. History is literally embedded in" them. "Even if certain histories are excluded" from TV shows, "they cannot be entirely silenced, because there will almost always be some piece of evidence" in the filming location itself "that we can use to challenge dominant historical narratives and recover hidden histories." Finally, focusing on locations can help us "rethink commonsense understandings of history and local geography and of the unequal relationships of power that sustain them."[39] TV shows intervene in their locations, helping to reproduce a sense of place that often affirms and naturalizes racial inequality. Textual methods alone can miss how media makers, together with state agencies, promote TV production as a boon to the communities where they film, thereby legitimating a place-based status quo, or new forms of "creative destruction" that build on the old.

State subsidies for entertainment commodities reproduce and widen place-based race and class inequalities. This is the conclusion, for example, of research about public subsidies for the building of expensive professional sports stadiums that dramatically remake urban space while redistributing wealth upwards. Although proponents of such subsidies claim they will produce economic growth, the scholarly consensus is that the gains represented by a professional sports franchise are small, without significant spill-over effects into other businesses and jobs. As Kevin J. Delaney and Rick Eckstein argue in *Public Dollars, Private Stadiums: The Battle over Building Sports Stadiums*, "using public dollars for private stadiums is another mechanism for transferring social resources from the not-so-powerful to the already powerful . . . (that is) likely to further polarize social inequality in the United States."[40] In his study of the St. Louis Rams in the book *How Racism Takes Place*, George Lipsitz adds that pro-sports subsidies reproduce and extend distinctly racialized forms of economic inequality.

> No one publicly recognized the contributions made by 45,473 children enrolled in the St. Louis city school system to the Ram's (2000 Super Bowl) victory. Eighty-five percent of these students were so poor that they qualified for federally subsidized lunches. Eighty percent of them were African American. They did not score touchdowns, make tackles, kick field goals, or intercept passes for the team. . . . (But) tax abatements for profitable businesses including the Rams football team deprived St. Louis children of seventeen million dollars annually in educational funding.
> . . . Students from low-income families lost access to educational dollars so that they could be spent subsidizing the profits of the millionaire owner of the Rams.[41]

The devaluing of Black education in favor of sports franchises is one way that racism happens, and another, I argue, is the diversion of public resources from poor Black, Latinx, and Indigenous people to film and TV producers. Subsidies for professional sports and Hollywood represent mutually reinforcing forms of racial capitalism in place where the lure of celebrity and overblown promises of economic growth legitimate inequality. Indeed, these two contexts, professional sports and television, provide practical, defining examples of racial capitalism and the production of entertainment commodities. As part of larger retreats from social spending, entertainment subsidies divert public resources from education, health care, and low-income housing to projects of racial capitalism in place.

Finally, throughout this book I argue that TV racial capitalism is a settler colonial formation. To be sure, TV content includes settler colonial representations of Indigenous peoples and lands, but to borrow from the title of a famous essay by Eve Tuck and Wayne Yang, TV settler colonialism is not a metaphor. Film studios in Southern California have long sat on Indigenous land, and TV settler colonialism has more recently been exported to Georgia and New Mexico, where soundstages, western ranches, and mixed-use residential developments also occupy Indigenous land. Hollywood-supported gentrification in poor places is thus part of the ongoing history of settler colonial appropriation and displacement.

Throughout *Producing Precarity*, then, I emphasize the place-based aspects of TV racial capitalism, especially in chapter 2's discussion of local warehouses and filming locations; chapter 3's discussion of TV production and local police and prisons; chapter 4's study of anti-colonial resistance in New Mexico and Georgia; chapter 5's analysis of TV shows that mediate local struggles over memorials to white supremacy; chapter 6's study of Tyler Perry's interventions into property and housing relations in Georgia; and chapter 7's analysis of TV and gentrification in East LA, Georgia, and New Mexico.

Watching TV Where It's Made

The title of this section describes my method of interpreting TV programs as expressions of racial capitalism in place. TV shows shot in poor locations contain disavowed critical knowledge about how government incentives reproduce racial inequality and, more broadly, about the historical preconditions for contemporary poverty. That knowledge remains invisible, however, so long as we view TV shows as representations isolated from their places of production. While there are excellent, content-based readings that decode ideological meanings in different TV shows and genres, such research can suffer from a sort of textual reductionism that ignores racial capitalism, labor, and the role of government in media production and distribution.[42] Conversely, while important research in film and TV production studies draw critical attention to screen industry labor relations with a focus on the exploitation and precarious positions of below-the-line workers, they generally neglect the symbolic significance of screen content, as well as the larger relations of racial capitalism surrounding and supporting screen industries, including low-wage, racialized, and gendered work in hotels, restaurants, janitorial services, and on farms.[43]

Drawing inspiration from Black, Chicanx, and Indigenous TV studies focused on the role of state power in media production, I instead attempt to bring together political-economic and textual analysis.[44] Which is to say I reconstruct a materialist history of TV production, demonstrating that the building of studios, combined with location shooting, contributes to the reproduction of particular kinds of unequal spaces, while also examining the implications of such material histories for understanding televisual texts. Although in their content TV series often repress or distort their locations to preserve their fictional worlds, the material conditions of their genesis nonetheless seep into their narratives and images, rewarding against-the-grain readings of televisual places.

TV programs filmed in poor places, I argue, reflect in displaced ways on their own mode of production and its basis in appropriation, exploitation, and incarceration. Several programs in this study represent TV production, and many more incorporate scenes of people watching TV, in ways that suggest self-reflexivity.

TV shows within TV shows, I argue, invite interpretations of the process of production. (See Figure 1.5.) Casting, narratives, and locations respond to local conditions by representing TV racial capitalism in disavowed forms. One method I use to connect TV programs and the immediate context of their making is the analysis of the architecture of TV production, starting with soundstages and other facilities. I also focus on preexisting buildings, as in the common use of industrial warehouses and factories for crime scenes, which, I argue, can be read as symbolic proxies for TV soundstages. Set design and dressing for homes

Figure 1.5: TV shows within TV shows: *The Curse*.

and apartments also represent interpretations of TV racial capitalism in place. The same can be said for the architectural styles of the real homes used in location shooting. Different architectural styles—"colonial," "Pueblo," "territorial," "mid-century modern," cheap motel, and even a van where a homeless person lives—represent the class hierarchies reproduced by TV racial capitalism in place.

DEI TV

Contemporary television is a part of the larger historical contexts of "diversity, equity, and inclusion," starting in the 1990s when governments, corporations, and universities established diversity directors and offices. "Diversity" overlapped with "multiculturalism," which have both been promoted as key to problem-solving and business success in a global marketplace. In recent history, DEI discourse and practice has been the object of far-right culture war attacks opposed to substantive transformations of dominant institutions. A significant body of scholarship has criticized corporate and educational diversity initiatives from an opposing perspective, arguing that they sometimes substitute formal equality and diverse representations for systemic transformation and material redistribution. Representational diversity promotes a belief in progress that is belied by realities of racialized exploitation and exclusion, hence helping to reproduce them. In a related way, many TV shows in this study combine diversity in casting with racial capitalism in their mode of production. What Melamed argues about institutional diversity in general is true of Hollywood in particular. DEI TV tries to "claim all differences (material, cultural, communal, and epistemological) for capital management," subordinating diversity to the aim of capital accumulation.[45]

This is supported by Kristen J. Warner's argument that diverse casting on TV in the early 2000s was partly a profit-making strategy prioritizing quantitative diversity over the qualitative diversity of cultural specificity and complex engagements with difference. Like other scholars of TV, Warner analyzes ideologies of post-feminist, post-racial colorblindness in TV shows such as *Grey's Anatomy* by showrunner Shonda Rhimes. Warner centers colorblind casting, in which characters are written as generic and racially unmarked (with whiteness as the assumed default) but then played by Black or Latinx actors. Colorblind casting is profitable because it represents racial progress to advocacy groups without alienating white audiences, since a diverse cast often takes the place of critical depictions of race in TV content. Building on Warner, I argue that diverse, colorblind casting is also a moneymaker because it helps legitimate TV production in poor places, obscuring how shows are paid for by diverting resources from Black, Latinx, and Indigenous people via state subsidies. At its most extreme, TV depicts police departments as committed to diversity, thereby propping up the police and prison televisual complex that helps manage and control the racialized, low-wage workers on which TV production in poor places depends (chapter 3).

DEI TV often defines positive, progressive forms of diversity in opposition to a dangerous Mexican presence. Racial capitalism in place is solidified, for example, by the incessant depiction of Mexican narco violence. In show after show, narcos are evil ghosts in the machine, working like racialized McGuffins to motivate narrative action without any content or context of their own. These reifying representations of Mexican violence have become clichés on crime TV, but they pop up across genres, even comedies. Given the dominance of white men in the TV industry, we can understand the fetishized figure of the Mexican narco as a legitimating foil for whiteness that is reproduced even by Black and Latinx creatives. One example is the show *Devious Maids*, with Eva Longoria as executive producer and a diverse writers' room (chapter 3). Despite or perhaps because of the show's narco plot, Longoria seemed to promote *Devious Maids* as a contribution to diversity that brought new stories of underrepresented people to the screen: "I've known so many amazing women in my life who are amazing," Longoria said. "And I said, 'You're saying these people don't have a story to tell—you're saying all they do is go and clean your homes.' They have big lives and complex lives."[46] She has also lauded the show's diverse cast: "It's the first and only show with an all-Latina leading cast. . . . I'm so excited. Five women from the Latina community! That, to me, was important—that we're reflected on television."[47] Positive representations of diversity however, go hand in hand with naturalized images of Mexican criminality and violence, in this case the members of the Gaviota cartel, which threatens to kidnap the daughter of one of the maid's employers. The narco as demonized figure for the Mexican immigrant poor represents "bad diversity," the defining foil for the "good diversity" showcased in *Devious Maids* and other programs. In contrast with the situations of the actual low-wage raced and gendered service workers who support the show's production, the central characters work for white employers in exploitation-free environments threatened by a Mexican underclass. Juxtaposing good and bad workers, *Devious Maids* serves a labor policing function, with implications for Black and Latinx workers in Georgia and beyond. As such, the show forwards a distinctly Latinx racial capitalism in place, whereby subsidies and representations profit Latinx elites from Los Angeles at the expense of Black and Latinx (including the undocumented) underclasses in Georgia.

If post-feminist, post racial TV shows and scholarship is informed by the Obama era, many more recent programs respond to the election of Donald Trump and the rise of the MAGA movement around 2015. Breaking with colorblind representations, shows such as *Atlanta, Gentefied, Being Mary Jane, Roswell, New Mexico, Vida, Watchmen*, and others present a post-colorblind world of racism and white nativism. This newer brand of more racially explicit shows is partly the result of increasing diversity among show creatives. Although they remain under-represented both in front of and behind the camera, the number of people of color, including women, working in TV has notably increased, as represented by many of the shows analyzed in this book.[48] Mary Beltrán argues that the success of Shonda Rimes's *Scandal*

(2012–2018) paved the way for other Black women and Latina TV show creators, including Cristela Alonzo, who created, co-wrote, starred in and co-executive produced the sitcom *Cristela* (ABC, 2014–2015); Gloria Calderón Kellett, co-creator of the Latinx reboot of *One Day at a Time*; and Tanya Saracho, who created and co-wrote *Vida*, a dramady about two Latinx sisters and gentrification (Starz, 2018–2020, chapter 7). I would add to the list of Latinx TV creators Linda Yvette Chávez and Marvin Lemus, who created another gentrification family drama, *Gentefied* (Netflix, 2020–2021, chapter 7). Contributors to the volume *Representations of Black Womanhood on Television: Being Mara Brock Akil* draw critical attention to the work of this successful creator and writer of *Being Mary Jane* (OWN, 2013–2019) and other shows (chapter 6). Meanwhile, the superhero show *Black Lightning* (2018–2021) was created by Brock Akil's husband, Salim Akil, who also directed many episodes (chapter 5). Donald Glover and Janine Nabers co-created *Swarm*, a limited series featuring a queer Black serial killer (chapter 4). Nabers was also a writer on Glover's surreal Black comedy *Atlanta* (FX, 2016–2022, chapter 4), and for the science fiction show about superheroes and white supremacy, *Watchmen* (HBO, 2019, chapter 5). Shows in my sample by Black creatives also include Misha Green's horror show exploring racism and segregation, *Lovecraft Country* (HBO, 2020, chapter 4); Ava DuVernay's queer Black teen superhero show *Naomi* (chapter 5); Jamey Giddens's political thriller *Ambitions* (OWN, 2019, chapter 6); Felicia D Henderson's HBCU drama *The Quad* (BET, 2017–2018, chapter 6); and Tyler Perry's White House thriller *The Oval* (BET, 2019–, chapter 6). Finally, a revealingly critical reflection on filming in poor places, the crime drama with a Cambodian immigrant woman protagonist titled *The Cleaning Lady* (chapter 3) was created by Chinese Canadian immigrant Miranda Kwok. In her study of Black comedy performers, Bambi Haggins demonstrates that writers and performers such as Eddie Murphey, Chris Rock, and Whoopi Goldberg have been effectively squeezed between the demands of white-dominated media industries and the desires of Black audiences for critical commentary, and the former often trumps the later.[49] Here I extend her analysis, arguing that TV creatives of color struggle with both industrial forces and the burden of representation.

Programs created by people of color receive significantly smaller budgets than those created by white people.[50] *The Quad* showrunner Henderson writes that the "least pleasant" part of the job is "the ongoing battle for an adequate budget." Even though by many measures the show was a success, it was constantly threatened with cancellation. She also reports struggling against assumptions that there were no qualified women of color directors, as well as the uncomprehending sexism and racism of network executives.[51] Henderson's experience is shared by many other Black and Latinx showrunners. As Robin R. Means Coleman and Andre M. Cavalcante write,

> While it is possible for producers and their creative teams to successfully write across racial and cultural lines, it is often the responsibility of Black showrunners to proffer Black imagery that runs counter to established industry canons. However, the influence of Black televisual image-makers . . . is limited, as they do not hold complete autonomy in resolving the split image (dividing conventional industrial representations and Black self-representation). Pressures from industry executives and the structural limitations of the televisual form continually present creative hurdles that obligate representational compromises, a dynamic that constitutes popular Black imagery as a site of continuous negotiation and redefinition.[52]

Despite these limits, the role of show creator is an important one since they are also often showrunners, writers, and directors, and they have significant influence over content and who else is hired to write and direct. As a result, the preceding shows employ a relatively diverse group of writers and directors, and in content they present under-represented perspectives and challenging stories, often referencing topical political issues such as police violence, detention and deportation, racism and sexism in the media, and white nationalism. However, the numerical diversity of behind-the-scenes TV workers does not necessarily result in better, more equitable representations. While not colorblind, in other words, progressive content in MAGA-era programs often hides the inequalities of TV production for local communities.

In *Hollywood Diversity Report 2023, Part 2: Television*, Ana-Christina Ramón, Michael Tran, and Darnell Hunt reach a sobering conclusion concerning Hollywood's DEI efforts:

> Though studios may have responded to pressure from advocacy groups with more diverse representation in front of the camera since we launched this report series a decade ago, these efforts have fallen short of meaningful inclusion because there have only been incremental increases in the shares of women and people of color as show creators, writers, and directors—critical behind-the-camera positions. After numerous proclamations of a commitment to increase diversity during the summer of 2020, just a couple of years later many studios and networks began cutting positions held primarily by people of color and eliminating pipeline initiatives and training programs. Some have canceled projects that never aired or soon after they aired and removed diverse films and television series from their digital platforms for cost-saving purposes. Indeed, this past summer, many of the Black women who had been hired in Hollywood diversity-related, leadership positions—some of which were created in the aftermath of the (George) Floyd murder (by police) and the nation's racial reckoning—were no longer in their positions due to "restructuring" or other reasons. This led many to wonder if industry, post-Floyd proclamations were just performative.[53]

In the context of state-subsidized TV production in poor places, we might wonder what exactly does Hollywood DEI perform? Many programs present seemingly liberal representations of race, gender, and sexuality that serve to disavow exploitation in their mode of productions. In contrast with contexts where investments in diversity in education and other institutions can lead to practical changes, in the TV contexts analyzed here representations of diversity take the place of substantive forms of equity and inclusion.

Although the two historical contexts are distinct in many ways, Robinson's discussion of silent films in *Forgeries of Memory and Meaning* helps us see how an industry of TV racial capitalism in place profits from anti-racist representations.[54] Writing about Black-made "uplift" or "race" movies (1910–1930), Robinson argues that although some scholars have read them as opposing mainstream anti-Black representations, most such works "exploited Black caricatures for the amusement of their largely Black audiences."[55] The promotion of comedies partly reflected the preferences of white exhibitors for caricatures of Black people. According to Robinson, during the 1920s, white businessmen, recognizing a lucrative investment, began to buy Black-owned theaters and build new ones in Black neighborhoods, further reinforcing the predominance of Black caricature.[56] Although some independent Black pictures, particularly those by Oscar Micheaux, opposed racial capitalism, however unevenly, "others delved deeper into the very construction of knowledge which collaborated with racism," including the Black middle class patrolling of race movies and the Black working-class "for signs of rebelliousness."[57] "The most daring of the race film producers," Robinson concludes, "seemed to have been contented with displays of bourgeois respectability and modest uplift themes" rather than "a profound challenge or radical critique of racial capitalism."[58] Appealing to Black audiences by incorporating Black people into movies, white capitalists and middle-class Black filmmakers sold a limited representational equity as a substitute for a broader social equity. Here, in his critical focus on bourgeois Black filmmakers, Robinson suggests that the other to the "Black radical tradition" is Black racial capitalism, an important claim that is lost when readers of Robinson focus only on whiteness.[59] Building on Robinson's insights, *The Costs of Making TV in Poor Places* elaborates theories of Black and Latinx TV racial capitalism in place to analyze how Black and Latinx creatives profit at the expense of poor and working-class Black, Latinx (including the undocumented), and Indigenous people where TV programs are produced.

Many shows in this study are aligned with progressive race politics yet leave unchallenged—and in fact often actively obscure—the political economy that makes their progressive representations possible. Via state tax incentive programs and other subsidies, progressive white, Black, and Latinx TV producers profit at the expense of poor people of color, even when such creatives and their shows remain marginal and vulnerable to corporate disposability. This is consistent with the history Robinson recounts but also distinct: a form of racial capitalism in place where TV programs siphon funds that could be spent on social welfare to instead pay for their critical narratives, often produced by Black and Latinx creatives. Media corporations in effect "launder" public money appropriated from Black, Latinx, and Indigenous people by converting it into DEI spectacles. TV programs reproduce racial capitalism in both their content and mode of production, especially when they seem to be at odds, the first making it harder to see the second.

TV Guide

The Costs of Making TV in Poor Places draws on over 70 TV shows, including a handful made in East Los Angeles and other places, but most were filmed in Georgia and New Mexico. I analyze a roughly equal number of shows shot in the two locations, although a few were filmed in both Georgia *and* New Mexico. Different chapters engage different scripted TV genres (crime, horror, science fiction, melodrama) streaming on multiple platforms.[60] Netflix accounts for the largest number of shows in my sample (12); followed by HBO (6); AMC and Fox (5); ABC, Amazon, BET, The CW, FX, OWN (4); Starz and USA (3); Disney, Epix, NBC, and Peacock (2); and finally, Here, Hulu, Showtime, Shudder, Sundance, and Syfy (1). I draw on programs across different platforms—broadcast, cable, and digital—although the last accounts for the largest number of shows and even the other two platforms make content available for streaming. Several popular, critically acclaimed "quality TV" programs such as *Breaking Bad*, *Lovecraft Country*, *The Walking Dead*, and *Watchmen* were filmed in New Mexico and Georgia, for example, and they have each inspired a significant body of scholarship. But I also focus critical attention on a host of other more obscure, less studied, yet nonetheless revealing productions. Except for a few outliers, the shows were mostly produced in the wake of Donald Trump's successful first campaign for US President, so from around 2015 to the present. Many take up topical issues of that time, including white nationalism; anti-Black police violence and the Black Lives Matter movement; migrant detention and family separation; and conflicts over Confederate monuments. In my efforts to interpret the social significance of TV content in terms of production context, I draw upon a range of discourses, paying particular attention to how Black, Latinx and Indigenous showrunners, writers, and actors discuss their work and its meaning.

While the focus here is on the domineering operations of TV racial capitalism in place, I also highlight examples that challenge the status quo. A handful of programs, I argue, depart from the representational politics of most others by connecting TV content to its context of production. *The Cleaning Lady* critically connects TV to regimes of immigrant detention and deportation; *Black Lighting* links its Atlanta location to struggles over Confederate Monuments. Finally, the performance of Indigenous actor Tamara Podemski as Joy Hawk, a queer Indigenous Sherriff in the New Mexico-made show *Outer Range*, draws into critical relief the settler colonial demands of DEI TV. Finally, I study local activists opposed to TV racial capitalism in place, including Black environmentalists, anti-police activists, Muscogee (Creek) water and forest protectors in Georgia, and anti-gentrification activists in East Los Angeles.

Chapter 2 presents a critical map of the architecture of TV production in Georgia and New Mexico centering on the physical locations of film studios, as well as filming locations in the surrounding locales, particularly old factories and warehouses. I argue that warehouse locations double for studio soundstages, and so many shows can be read as revealing, albeit displaced and disavowed, representations of TV production and its consequences for poor places: the appropriation of life-sustaining resources; racialized and gendered hierarchies of labor exploitation; and colonial settlements. This chapter analyzes the material implications of state-subsidized TV production in part by interpreting a range of shows, while subsequent chapters focus on individual genres.

Crime dramas, the most popular TV genre, are the subject of chapter 3, where I argue that many shot in Georgia and New Mexico depict the police as champions of diversity. While they sometimes represent bad cops, on balance TV police look good, particularly when contrasted with Hollywood's favorite villains, Mexican narcos. Such representations are symbolic expressions of the police and prison televisual complex linking TV production and the local criminal justice system. Incentive programs effectively subsidize the practical alliances articulating police, prisons, and TV production. TV racial capitalism in place, in other words, is also carceral, both in terms of representation and the practices of TV production. This chapter engages shows about the police (*In Plain Sight, Killer Women, Deputy, Walker: Independence, Will Trent*, and *Teenage Bounty Hunters*) and about criminals (*Breaking Bad, Better Call Saul, Ozark, Devious Maids*, and *The Cleaning Lady*).

In chapter 4, I argue that horror shows employ generic conventions to construct a DEI gaze that disavows local conditions of production. A trio of programs focusing on serial killers—*The Unsettling*, an episode of *Atlanta*, and *Swarm*—depict the horrors of foster care as a legitimating foil for TV production as a better use of state funds. The shows effectively suggest that TV makes a greater contribution to diversity than social spending. A group of supernatural horror shows with diverse casts and anti-racist narratives—*Midnight, Texas, Sleepy Hollow, Preacher*, and *Lovecraft Country*—aestheticize and rationalize forms of disposability characteristic of racial capitalism in place: the disposability of Black women workers, communities of color, and the environment.[61] *Atlanta* and *Lovecraft Country* in particular invite investigation into the violence that accompanies local location shooting.

Chapter 5 focuses on science fiction TV. Georgia and New Mexico host extensive science fiction productions, including many DC and Marvel films and TV programs. Atlanta-shot shows about Black superheroes include *Watchmen* (HBO, 2019), *Raising Dion* (Netflix, 2019–2022), *Naomi* (The CW, 2022), and *Black Lightning* (The CW, 2018–2021), where Black superheroes fight evil, especially anti-Black racism. Partly because of the history of Area 51 and stories of alien sightings, New Mexico has attracted TV shows depicting alien phenomena, as in *Outer Range* (Amazon, 2022–) and *Roswell, New Mexico* (The CW, 2019–2022), two science fiction westerns shot in Las Vegas and Santa Fe respectively. Both kinds of science fiction employ diverse casts and behind-the-scenes talent, and they critically engage topical political issues like police violence, white nationalism, Confederate memorials, settler colonialism, and immigration enforcement. Hence, while set in speculative worlds, science fiction shows remain tethered to the real world. Which is also to say that such shows can't escape the conditions of their own production and thus represent in displaced and distorted forms the racial capitalism in place in which they participate.

The genre of Atlanta-made melodramas analyzed in chapter 6 represents Black racial capitalism in the form of elite Black institutions and figures in religion, education, law, politics, business, and the media. In addition to *Greenleaf*, *The Quad* (BET, 2017–2018) is set at an HBCU; the main characters in *Ambitions* (OWN, 2019) are lawyers, Atlanta politicians, and CEOs; *The Oval* (BET, 2019–) is about a fictional mixed first family; and finally, *Being Mary Jane* (OWN, 2013–2019) focuses on a Black TV news reporter. From the vantage of such settings, Atlanta-shot shows present complex, critical representations of anti-Black racism and sexism in the MAGA era. But they focalize their critical representations of anti-Black racism and police violence through wealthy Black characters. The programs in this chapter represent working-class Black life from middle- and upper-class Black perspectives. In the form of public subsidies for film and TV production, the upward redistribution of wealth from poor Black people to more privileged Black media makers pays for depictions of Black wealth. The common thread among the different shows is a relative support for the police, in keeping with the practical alliances among TV production, police, and prison analyzed in chapter 3.

Chapter 7 analyzes TV shows that represent gentrification. While focusing largely on Georgia and New Mexico, here I also analyze filming locations in the working-class Latinx area of East Los Angeles. To compete with other states and cities, the state of California and City of Los Angeles provide a host of subsidies to film and TV producers. In the remainder of the chapter, I thus present a comparative study of gentrification shows in New Mexico (*The Curse*) and the Boyle Heights neighborhood of East Los Angeles (*Falling for Angeles*, *Vida*, and *Gentefied*). All four programs are complicit in the kinds of gentrification they depict, but the New Mexican setting of *The Curse* sheds critical light on the representation of gentrification in the others. The juxtaposition of New Mexico and Los Angeles settings invites investigation into the occluded history and ongoing reality of film and TV production in greater Los Angeles as a settler colonial project, not only in terms of representation but also via the occupation of Indigenous land.

Finale

In both their content and production, state-subsidized TV shows shot in poor places exemplify racial capitalism in place, where capitalism takes local, racialized directions. Originating in colonial and plantation histories, racial capitalism in Georgia, New Mexico, and other precarious locations has morphed into film and TV production. Despite projections of economic growth, the regional benefits of state subsidies seem small relative to the costs of diverted social spending. Screen industries employ some local workers in TV production, but mostly depend on racialized and gendered low-wage labor in the surrounding economy. New studios and their mixed-use housing developments results in higher housing costs that displace poor people and encourages increased policing to protect private property. Profiting at the expense of Indigenous people and poor people of color, Hollywood helps drive both the slow violence of social disinvestment and gentrification, and the swift violence of police and prisons. With their combination of gentrification and policing on Indigenous lands, TV studios have become powerful players in the ongoing history of settler colonialism. Many shows made in such contexts represent progressive race politics that serve to displace from view and critical reflection the racial capitalism in place that makes such representations possible. Progressive racial representations can thus serve a repressive end, as suggested by the many shows that depict the police as champions of diversity, equity, and inclusion.

Get Shorty crystalizes the violence of TV production obscured by diverse TV content. The show's main character, Miles Daly, works as a hit man for Amara the Guatemalan drug queen. When a screenwriter racks up gambling debts to finance a film, Amara sends Miles and his partner to Hollywood to collect. They kill the screenwriter, but Miles takes his bloody script to Amara and pitches the plan of financing the picture to launder drug money. Using violence and intimidation, Miles secures a director and a studio, and successfully produces *The Admiral's Mistress*. In the final episode of season 1, an executive for MGM (the studio that in real life produced *Get Shorty*) offers him another producing job based on his reputation for getting things done by being "very fucking persuasive." When asked by an interviewer how accurate was the show's depiction of Hollywood, actor Chris O'Dowd, who plays Miles, suggested that gangsters and TV producers aren't that different: "It's an industry awash with charlatans and drowning in money—some of it from good places. But even if you're wearing a shirt and tie, it doesn't necessarily mean that you're a good guy. Particularly so much money from different sources coming into the industry that nobody really questions that much, and because it doesn't suit them to know the truth. Maybe some of the real heavy-handed physical stuff (torture, murder) doesn't happen so much, but there are definitely touches of it."[62] *TV in Poor Places* follows the money that nobody questions because it doesn't suit them to know the truth. But that trail leads not to TV narcos, but from poor people to Hollywood. The ubiquitous TV and film figure of the Latin American narco in this reading is a displaced self-representation of the legal, even police protected "crime" of TV production in poor places.

2

Poor Locations

In *Out of Stock: The Warehouse in the History of Capitalism*, Dara Orenstein suggests that warehouse settings are a defining feature of popular crime dramas, furnishing "a narrative container for murder and mayhem."[1] The warehouse crime scene is certainly a widespread cliché, but its use in TV programs shot in poor places suggests local meanings. Shocking acts of warehouse violence, for example, jumpstart the narrative of *Ozark* (Netflix, 2017–2022). Set in Missouri, *Ozark* was filmed at locations on the outskirts of Atlanta and on the soundstages of nearby Eagle Rock Studios. In the first episode, Mexican drug lord Camino Del Rio (Esai Morales) and his men murder several people inside a trucking company warehouse, forcing the show's money laundering protagonist, Marty Byrd (Jason Bateman), to beg for his life. Some of Eagle Rock Studios' soundstages were previously used as a Kraft Foods warehouse, and the studio's owner, Eagle Rock Distributing Company, distributes alcohol.[2] Eagle Rock includes multiple truck loading docks, making its soundstages resemble the trucking company setting of cartel violence in *Ozark*. This example and many others suggest the possibility of interpreting warehouse settings as indirect representations of the material spaces of TV production.

Most films and TV shows employing warehouse settings are themselves partly shot in warehouse spaces. State subsidies in Georgia and New Mexico have encouraged local film and TV studio building sprees. Taking advantage of partnerships with state and city governments that subsidize the purchase or rental of land, buildings, and other infrastructure, real estate developers have opened numerous studios with large warehouse soundstages. Several of the new studios are in older industrial warehouses that have been refurbished for filming. Studio warehouses are also increasingly part of larger, mixed-use developments including retail businesses and residential communities. TV representations of warehouses and related spaces can be read as interpretations of the recent history of such developments, interpretations that guide my critical account of TV racial capitalism in place. Film and TV studios in New Mexico and Georgia further anchor extensive local location shooting in materially and symbolically significant places such as factories, prisons, and Native American reservations. In many contemporary TV shows, representations of warehouses and related settings serve as containers for narratives symbolizing the material consequences of state media subsidies and the regional forms of racial capitalism to which they contribute: the appropriation of life-sustaining resources, racialized and gendered hierarchies of labor exploitation, and colonial settlements.

TV Warehouses

In addition to filming on local warehouse soundstages, TV producers in Georgia and New Mexico often use closed or abandoned industrial warehouses as filming locations. During its 12-year run, for instance, *The Walking Dead* (AMC, 2010–2022) prominently featured decaying warehouses. Scenes of the Woodbury Arena, where members of a survivor community fight zombies for entertainment in season 3, were filmed next to an old water tank warehouse in Newnan, Georgia, while the location for the survivor community called "Terminus" in seasons 4 and 5 was a historic railyard with a large warehouse on the outskirts of Atlanta.

Railroad warehouses are particularly popular settings for crime and violence. Albuquerque's distinctive old railyards are featured in numerous shows, including well-known "quality TV" but also a host of more obscure programs. The railyard buildings were used for exterior shots as Jesse Pinkman and hitman Mike Ehrmantraut pick up a drug money drop, and as a stand-in for Walter White's meth lab in promotional material for season 5 of *Breaking Bad* (AMC, 2008). In the Albuquerque-set show about witness protection titled *In Plain Sight* (USA, 2008–2012), the railyards were used for Boston and Detroit crime scenes (S3E1, S4E7). In season 3 of *Better Call Saul* (AMC, 2017), Ehrmantraut meets a criminal associate there. In the first episode of *Get Shorty* (Epix, 2017) the location serves as an out-of-the-way place for two drug cartel hitmen to clean a bloody car and hide a corpse. In the surreal Texas-set crime drama *Briarpatch* (USA, 2020), the railyards represent an old slaughterhouse and

Figure 2.1: Lucifer and a drug lord named El Jefe plan a murder inside an Albuquerque Rail Yards building in *The Messengers*.

the site of clandestine meetings and murders. Finally, in *The Cleaning Lady* (Fox, 2022–), two drug smuggling informants and their FBI handler meet at the Albuquerque Rail Yards (S2E12).

The railyards have also been employed in science fiction and horror shows. *The Lost Room* (Syfy Channel, 2006) uses a railyard warehouse for a scene in which a villain offers to return a police detective's daughter in exchange for a strange key that opens doors into other times and places; the young girl, however, mysteriously disappears behind a door in the warehouse. In the premiere episode of *Terminator: The Sarah Connor Chronicles* (Fox, 2008–2009), killer cyborgs from the future pursue the title character, who hides in a glass railyard building. The central villain in the YA zombie apocalypse series *Daybreak* (Netflix, 2019) lives in one of the railyard warehouses. In the supernatural show *Midnight, Texas* (NBC, 2017–2018), a railyard warehouse serves as the location for a battle between a psychic and a magician (S1E5, 2017). In another supernatural thriller titled *The Messengers* (CW, 2015), Lucifer meets a drug lord named El Jefe in the railyards to plan a murder (S1E3). (See Figure 2.1.)

The railyards have even seemingly inspired the use of similar settings for shows filmed elsewhere. With its decaying metal structures and wall of green windows, the Georgia set for the Savior's community led by the villain Negan on *The Walking Dead* seems to have been self-consciously modeled on the buildings in Albuquerque.

The railyards are jointly managed by the city of Albuquerque and Netflix Studios, making the location a sort of public subsidy for the streaming giant, since it is cheaper to film in preexisting structures than to construct comparable new sets. In addition to being employed as a practical location, the railyards are also used like a conventional soundstage. According to Jason Strykowski, "thanks to the sheer size of the structures, film crews can use the interiors as fully functional studios by rigging green screens and constructing full-scale sets." He goes on to note that for *The Avengers*, Marvel Studios constructed scale sets of blocks in New York and Kolkata, India inside railyard buildings.[3] We can think of the railyard as a practical and symbolic double for the new film and TV studios in New Mexico, while TV shows filmed in those warehouses provide critical insights into the ideas and practices defining publicly subsidized media making in poor places. (See Figure 2.2.)

With the Albuquerque Rail Yards, TV producers fetishize urban decay and, as I argue in chapter 4, romanticize racialized poverty. The use of the railyards, however, also suggests that film and TV have triumphantly superseded earlier, Fordist industries. With their steel frames and distinctive glass roofs and glass curtain walls, the railyard buildings

Figure 2.2: The Albuquerque Rail Yards.

were on the cutting edge of industrial architecture in the early twentieth century. Their neoclassical style represented industrial efficiency, and as the largest employer in the city until they closed in the 1930s, the railyard buildings "were viewed with community pride as signs of progress and prosperity."[4] Today, film and TV boosters claim that state subsidies bring a new industry to town, with high-tech infrastructure and local jobs. Although as I noted in the introduction, job gains are small relative to their cost, the prospect of employment and economic development have turned TV and film production into emblems of the kind of progress and prosperity formerly represented by the railyards.

A similar progressive narrative about film and TV production in Georgia is suggested by Atlanta's Third Rail Studios, where the final season of *Ozark* was shot. It's built on the site of GM's old Doraville assembly plant and named for the three rail lines that previously served it. The name also refers to a primary selling point for the studio on its website: proximity to a light rail transit terminus linking it to Atlanta neighborhoods and two airports.[5] But in implicit contrast with the older industries it replaces, Third Rail is advertised as "a purpose-built film studio in Atlanta, GA on a beautiful, secure campus consisting of state-of-the-art, soundproof stages, ample mill and support space, and spacious, boutique production offices." Third Rail Studios was recently purchased by Gray Television, which is in the process of building an even larger studio called "Assembly Atlanta." Michael Hahn, the Los Angeles developer behind Third Rail Studios, has also established Electric Owl Studios, likewise adjacent to a rail terminus.[6] "Electric Owl" is the early twentieth century nickname for a night train conductor, but again in implicit contrast with Fordist industries, the studio promotes its "green" and "sustainable" practices.[7] The use of outmoded warehouses for film and TV locations supports such implicit contrasts, positing old railyards and auto plants as legitimating foils for contemporary screen industries. In this regard, film studios in Atlanta and Albuquerque resemble de-industrial Belfast, where *Game of Thrones* and other shows were filmed. According to Ipek A. Celik Rappas, "Screen industries and their promotion in Belfast shows that while manufacturing is waning, its discourse of social welfare, hard labor, and craftsmanship transfers itself to creative industries that then justify themselves through the claim to inherit traditional industries' economic strength, job opportunities, and work ethics."[8]

Similar contemporary representations in Georgia and New Mexico rearticulate early twentieth century discourse in which film studios represent models for modern industry in the service of nationalism and racial hierarchies. Early twentieth century Japanese commentators, for example, saw glass film studios as icons of national modernization. Japanese studios, Diane Wei Lewis concludes, were "a small part of a rich global cultural field in which new visual and spatial idioms were used to celebrate industrialism, stimulate consumption, structure national identity, and shape attitudes toward racial and ethnic 'others,' reproducing (and extending) the regulatory mechanisms of the modern nation-state while justifying imperialist expansion."[9] Similarly, Rielle Navitski argues that in 1920s Brazil, with its history of slavery and a large Black population, newspaper and magazine writers promoted film studios as a means of demonstrating that Brazil was a modern, industrialized, and hence white nation. In the controlled environment of soundstages, blackness could be excluded, and whiteness centered, while the technical mastery required to make studio films symbolically whitened Brazilian films.[10] Contemporary Georgia and New Mexico suggest comparison to historic Japan and Brazil as relatively peripheral places in their respective world systems.

More than just modern industrial marvels, however, several new studios also anchor affluent communities, heirs to the "broadcast cities" analyzed by Lynn Spigel. Starting in the 1930s, Radio City and other TV broadcasting buildings "used the city concept," including mixed-use commercial spaces and proximity to transportation lines, "to mark their value as urbane, civic and, above all, *modern* environments for addressing and gathering publics through the wires."[11] Similarly, contemporary studios are promoted as modern, mixed-use real estate developments which, I argue in chapter 7, constitute forces of gentrification. Even when they don't include residential communities, studios publicize their proximity to affluence. Third Rail advertises the short light rail ride from the studio to the white majority neighborhood of Bucktown, "home to Atlanta's finest hotels, shopping, dining and residential neighborhoods."[12] The Atlanta studio where episodes of *Stranger Things* and *Saving Dion*

36 | POOR LOCATIONS

Figure 2.3: Shadowbox Studios, bordered by waterways, forests, and the Georgia Department of Corrections or "Cop City," in an area over 70% Black.

are filmed, EUE/Screen Gems (rebranded as Cinespace Atlanta) promotes its five-minute distance from "downtown luxury hotels."[13] Similarly, the website for Shadowbox Studios (formerly Blackhall Studios), a set of 9 soundstages made from refurbished commercial warehouses on over 100 acres of land, proclaims that its "strategic location places talent and crew only minutes away from the centers of city life—near Midtown, East Atlanta and Inman Park neighborhoods and Atlanta's luxury accommodations such as the Ritz Carlton, St. Regis and the Four Seasons. (See Figure 2.3.) The studio is readily accessible to the city's award-winning, celebrity chef-driven restaurants and high-end shopping destinations."[14] Shadowbox/Black Hall is close to the increasingly white and gentrified neighborhood of Midtown, and the exclusive neighborhood of Inman Park, with a history of restrictive covenants.[15]

Meanwhile, in New Mexico, Santa Fe Studios compares its film offices for rent, with their kitchenettes and balconies, to a Four Seasons Hotel room.[16] Indeed, the facilities are built in the modern "Pueblo Style" popular among luxury resorts in the area.

In addition to tax subsidies for the film and TV producers who rent them, many studios are public/private partnerships that divert money from schools and low-income housing to real estate developers. The Netflix expansion in Albuquerque, for example, is partly funded by

$20 million in state and city Local Economic Development Act (LEDA) funds. Albuquerque has also issued an Industrial Revenue Bond (IRB) to reduce studio property and other taxes over a 20-year period.[17] Finally, the city has spent $8 million in transportation taxes on a new freeway interchange to improve access to the Mesa del Sol development connected to the Netflix studio.[18] About an hour north, the construction of Santa Fe Studios was funded by a $10 million state grant and a tax-free IRB.[19]

In Georgia, Tyler Perry purchased an old army base from the state at a 60% discount, buying land appraised at $73 million for $30 million in 2015. The state also subsidized the building of a road adjacent to the studio.[20] Similarly, in 2022 the Decide Dekalb Development Authority approved a $34 million tax break and a $34 million "infrastructure credit" for Shadowbox/Blackhall.[21] The studios' developer, Ryan Millsap, reached an agreement with the city to swap a parcel of land he owns for Intrenchment Creek Park, public land next to the studio and alongside the South River. As entertainment reporter Jenny Fuster writes, the exchange was met by protests from Black environmental activists:

> The South River Watershed Alliance [SRWA] remains skeptical of both Blackhall's promises and the county's assurances that the proposed park would be a fair environmental tradeoff for what they estimate will be a loss of 3,000 trees clearcut by all of Blackhall's expansion efforts. [SRWA President] Jacqueline Echols said that Atlanta's tree canopy is among the best safeguards Atlanta has against the rising temperatures of climate change; and once it is cut down, there's no restoring it. "This swap would set a dangerous precedent of giving up parcels of public land to private businesses," she said. "These forests are the last bit of truly natural space we have in Atlanta and rather than care for them and take care of the economic opportunities those forests give us, our county is giving it away to make way for more warehouses, which really is what these studios are."[22]

Echols goes on to emphasize the environmental injustice of the deal. The city government funnels sewage to the South River, which she notes runs through predominantly Black, low-income neighborhoods. Although the nearby Chattahoochee River, which runs through wealthier neighborhoods, has received significant environmental aid from local officials, the SRWA's demands for funds to clean up the South River, one of the most polluted in the country, have not been heeded. (See Figure 2.4.) "The City of Atlanta and DeKalb County have not invested in protecting the environment and public health of the East side of the city and in the South River. All of the industrial development historically has been here. These new plans by the county to welcome in Hollywood is just the latest phase of that."[23] Based on the studio's past environmental record, Echols is right to be concerned, since she discovered a large toxic tire dump on studio property.[24]

The land swap was briefly delayed by a 2021 SRWA lawsuit, but in the meantime, Millsap barricaded the entrance to the park.[25] In response, a group of young activists identifying as "forest defenders" set up camp there, protesting both the land swap and studio expansion as well as the building of "Cop City," an adjacent police training camp that also endangers water and forests. According to a local Fox News report, an anonymous protester claimed credit for a quickly extinguished fire at Blackhall/Shadowbox, writing in a blog post, "May this be a warning to them, a small taste of what's to come if they attempt to expand south of the river . . . We don't like movies. We don't like screens. We are in the real world: Unseen to the Hypno-Dystopic Civilization around us. Somewhere among shadow and tree."[26] One of the protestors' signs reads "#StopCopCity" and "#NoHollywoodDystopia."[27]

The diversion of public resources to gentrifying film and TV studios that threaten forests and waterways in Black communities critically illuminates TV scenes of warehouse violence. In *The Lost Room*, a police officer wields a mysterious microwave weapon to propel a criminal through railyard windows to his death. Similarly, *Midnight, Texas* employs the railyards to stage a fatal battle between the show's psychic protagonist, Manfred, and a villain named Hightower who uses telekinetic power to break windows and propel shards of glass at his enemy before ultimately turning the glass weapons on himself in an act of suicide. One structural feature of the railyard buildings that made them good for working on trains also makes them well suited for filming TV shows: their large, continuous glass curtain walls. Glass curtains enable both train mechanics and media makers to employ natural light in the day. At night the glass walls are excellent for low lit scenes and shadowy, mysterious effects. The railyard glass is thus also suggestive of film lenses and filters. By weaponizing the railyard glass shards, *The Lost Room* and

Figure 2.4: "Stop the Swap" poster opposing the land exchange between the city of Atlanta and Shadowbox/Black Hall Studios, www.stoptheswap.org.

Midnight, Texas indirectly implicate the material structures of TV production in death and destruction. Warehouse locations and their cognate spaces represent in displaced forms the violence of TV made in poor places. Consciously or not, TV show warehouses suggest the material costs and consequences of TV racial capitalism in place.

Diversion and Consumption of Life-Sustaining Resources

Writing about the cycle of 1970s horror films featuring cannibals, Robin Wood argues that "cannibalism represents the ultimate in possessiveness, hence the logical end of human relations under capitalism."[28] Here I argue something similar but more specific about TV cannibals as displaced depictions of Hollywood's consumption of potentially life-sustaining public resources. State and municipal subsidies for TV producers and local studios divert money from housing, health care, and education for people of color and Indigenous people in poor states. TV shows indirectly register this situation by employing studio warehouses as locations for sensational depictions of violent consumption. Characters following old rail lines in *The Walking Dead* (AMC, 2010–2022), for example, encounter numerous signs directing survivors to the "sanctuary" of the former train station called "Terminus." When they arrive, however, they are greeted by a community of cannibals who kill and butcher the unsuspecting in an old warehouse with glass windows like the Albuquerque Rail Yards. The settlement's motto is "You're the butcher, or you're the cattle." A second Terminus warehouse stores the possessions stripped from victims, including clothing, jewelry, and stuffed animals (S5E1). In the local context of TV production, this seems like a sensational allegory about the gap between the advertised benefits of the film industry in Georgia and its life-diminishing realities.

Similarly, set in Glendale, California, but filmed in Albuquerque, *Daybreak*, a Netflix dramedy about teen survivors of a zombie apocalypse, uses an Albuquerque Rail Yards warehouse as the setting where a high school principal named Burr (Matthew Broderick) butchers and cooks his students. The show combines three elements of TV racial capitalism in place. First, *appropriation*: Burr claims to want to help kids but then eats them, a macabre interpretation of the economic aid TV producers and studios promise local communities as they consume public resources that could otherwise go to support them. Second, *incarceration*: *Daybreak*'s producers built an elaborate set inside a cavernous rail yard building that resembles a migrant detention center, projecting a sort of speculative vision of the symbiotic relationship between TV production and local carceral institutions. (See chapter 3.) And finally, *exploitation*: the prison is guarded by international students Burr has forced into the job.

Recalling the Terminus motto about butchers and cattle, studio warehouses are also represented by slaughterhouses. The evil Quincannon Meat and Power Company in season 1 of the supernatural horror show filmed in New Mexico *Preacher* (AMC, 2016–2019) is headquartered in a massive warehouse where cattle are butchered. The town of Annville is desperately poor, and the presence of the corporate slaughterhouse doesn't help; on the contrary, the company's CEO, Odin Quincannon, leads a militia attack on the town. Ultimately, however, the accumulation of methane in his warehouse creates a massive explosion that destroys Annville and kills all its inhabitants. *Preacher* seems like a parody of the new studios in New Mexico, whose proponents make inflated promises of economic growth while consuming public resources that could instead sustain local life. *Briarpatch* (USA, 2020), another New Mexico-made series, uses an Albuquerque Rail Yards warehouse to represent an old slaughterhouse in "Packing Town," a poor, Mexican part of town where the carnage of the show's finale takes place, leaving four characters dead (S1E5 and E10). Read as a rendering of its mode of production, *Briarpatch* suggests that rather than supporting local livelihoods, studio warehouses undermine them by consuming resources that could instead pay for education, housing, and health care.

While in the previous examples the connections are implicit, *Watchmen* explicitly links a slaughterhouse to the process of making media. The episode titled "This Extraordinary Being" (S1E6), set in 1930s New York but filmed in Georgia, presents the story of a Black police officer, Will Reeves, who becomes the avenging, anti-racist superhero "Hooded Justice" after he attempts to arrest a white supremacist named Fred for firebombing a Jewish deli. In response, Reeves's white fellow officers threaten to lynch him. Reeves subsequently follows two Klansmen to a large brick warehouse for "F.T. and Sons Very Fine Meats," a business owned by firebombing Fred, who also owns a market in Queens. (These details suggest a reference to Fred Trump, Donald Trump's father, who also owned a market in Queens and who was reportedly arrested at a New York Klan rally in the late 1920s.) Reeves dons his hood and walks into the warehouse, and in a combination of shots mostly from his point of view, the camera follows him down a hall with slabs of beef hanging on hooks and into a large room where he finds several Klansmen, some in police uniform, packing film projectors into wooden crates labeled "Allied Projection Co." After shooting them all, Reeves sees a flashing red lightbulb over a door, which he enters to discover a soundstage, where a Klansman in a police uniform is recording a subliminal racist soundtrack for a Hollywood film that leads Black viewers to violently attack each other. *Watchmen* casts a critical gaze on Hollywood racism and its effects that, inadvertently or not, implicates its own process of production. While expressly anti-racist in its content, to the extent that the show and the studios where it was filmed consumed resources that could have been devoted to social welfare, *Watchmen* effectively undermines poor Black lives.

Labor Exploitation

In addition to state subsidies, studios and TV producers benefit from the larger conditions of racial capitalism in poor places, including gendered and racialized low-wage labor not directly in TV production, where even below the line jobs go to relatively more privileged workers, but in the surrounding labor market.[29] This mode of racial capitalism is particularly acute in New Mexico, one of the poorest states in the union, economically dependent on tourism and hence jobs in food service, hotels, landscaping, janitorial services, and security. In its 2020 report approving a $7 million LEDA subsidy for Netflix's Albuquerque Studio, the Albuquerque Development Commission noted that it "benefits hotels, food vendors and caterers, restaurants, dry cleaners, (and) local security companies," businesses that largely employ low-wage workers.[30] Luxury hotels in Albuquerque's gentrifying Sawmill District have experienced "huge growth" by housing and feeding visiting film and TV crews. Although subsidies generate relatively few local jobs, many of those it does support are service positions in which Latinx and Indigenous women are overrepresented.[31]

TV shows made in Georgia and New Mexico represent the low-wage Black, Indigenous, and Latinx workers on whom the industry depends, such as office and location cleaners, laundry workers, restaurant workers, convenience store cashiers, janitors, hotel staff, and Airbnb cleaners. As if visualizing the hierarchies of racial capitalism in TV production, such workers are mostly minor characters, known only by their first names, including "Lucy," a Denny's waitress who serves Walter White; "Cara," cashier at the Big Chief gas station who lets Jessie Pinkman pay with meth; and "Fran," the Loyola's Family Restaurant waitress in both *Breaking Bad* and *Better Call Saul*. More often, however, they are unnamed figures in the background, played by anonymous extras, including people who work at the real-world locations where shows are shot. Hotel and restaurant workers also appear on the margins of the frame, only partly visible. Sometimes only the black and brown feminine hands of servers are seen as they deliver plates and top off coffee cups, visually representing racial capitalism's tendency to reduce women of color to working hands.

TV's dependence on low-wage racialized and gendered labor is suggested by *Devious Maids*, set in Beverly Hills but filmed in the wealthy Bucktown neighborhood of Atlanta. In addition to employing local workers to clean up sets and locations, as analyzed in chapter 1, the cast and crew of *Devious Maids* extensively patronized local restaurants and advertised them to viewers. In a series of videos promoting film and TV production in Georgia made by the Georgia Film Office, program producers extensively praise Atlanta's food in between shots of working hands preparing food. In the video "Devious Maids: Three Seasons in Atlanta," producer/director David Warren praises the food on set and then the camera cuts to a scene of a Latina making sandwiches while Dianna Ramirez, who plays maid Rosie Falta, exclaims "the food is amazing."[32] In another video titled "Devious Maids Stars on Southern Hospitality and Food," Rebecca Wisocky, who plays wealthy socialite Evelyn Powell, says "We as a cast and as a *Devious Maids* family, we spend a lot of time in the restaurants of Atlanta." This is followed by scenes in a restaurant, including with closeups of hands preparing dishes as Judy Reyes (who plays maid Zoila Diaz) explains "My favorite thing to do in Atlanta is eat, shop, eat, eat, eat."[33] Finally, in "Devious Maids Loves the ATL! (It's Mutual)," Ana Ortiz (who plays maid Marisol Suarez) confesses, "I think

I've eaten my way through Atlanta."[34] Together, these state-produced advertisements for filming in Atlanta suggests that among the promised new jobs, many are in low-wage food service positions.

In a variety of ways, *Devious Maids* seemingly reflects on its own production context, including its dependence on low-wage racialized and gendered service labor. One of the maids, Carmen, works for a popular recording artist, while another, Rosie, works for film and TV stars. A third maid, Marisol, writes a best-selling expose of working for the rich, *Coming Clean: My Year Undercover as a Beverly Hills Maid*, which is optioned for a film of the same title in the show's final season. Season 4 begins on a soundstage where *Coming Clean* is being shot, with Eva Longoria playing the role of Flora, the maid murdered in the pilot for *Devious Maids*. Marisol is hired to write the TV movie within the TV show, but she's fired after clashing with a sexist director. Meanwhile, Carmen unsuccessfully auditions for the film role based on her, but it is implied that she is viewed as too old for the part. While likely inspired by the experiences of Latinx actors and creatives, including executive producer and script consultant Eva Longoria, Hollywood's appropriation of the maids' work and lives in the show's narrative allegorizes TV racial capitalism in place. The disposability of the maids is reinforced by the show's body count: out of the 10 people murdered on the show, 6 were Latinx, including 2 maids.

The Cleaning Lady also indirectly suggests that TV production is based in low-wage service work. The program is about an undocumented Cambodian woman named Thony De La Rosa (Élodie Yung) who migrates from the Philippines to Las Vegas to secure medical treatment for her son. Several warehouse scenes were filmed at the Journal Studio Center in Albuquerque. Episode 1 begins with Thony and her sister-in-law working as cleaners in a hotel before shifting to Thony's job cleaning toilets in a warehouse where she accidently witnesses a weapons dealer murder an underground fight promoter. To avoid being killed herself, she offers to clean the murder scene. In episode 2, she is coerced into cleaning up after a second fatal warehouse shootout between arms sellers and buyers. This second warehouse belongs to a trucking company that serves as a front for weapons dealing; with its multiple loading docks it resembles a studio soundstage. Finally, in episode 7,

Figure 2.5: Thony prepares to clean a warehouse murder scene in *The Cleaning Lady*.

Thony is forced to clean up a third murder in a warehouse storing food items. (See Figure 2.5.) *The Cleaning Lady* is thus a sensational rendering of the gendered and racialized, low-wage jobs promoted by film and TV production.

The articulation of warehouses and cleaning labor was anticipated by *Breaking Bad*, where drug lord Gus Fring builds a meth lab beneath an industrial laundry facility called "Lavandería Brillante." The lab's entrance is hidden behind an industrial washing machine in a large warehouse. Wearing matching grey shirts, the laundry workers appear to be Latinx. Across multiple episodes they are filmed punching a time clock, loading and unloading giant washers and dryers, and feeding linens into large pressing and folding machines (S3E6, S4E6 and E12). The extras playing the laundry workers were employees of the real-life laundry company where *Breaking Bad* was filmed.[35] In one episode, Walter White interrupts three women sorting laundry and pressures them to clean the underground lab. He speaks in broken Spanish and offers them $50 dollars each, "the universal language, *dinero*." As White drinks coffee and raises his cup in a toast directed at a security camera, the women clean chemistry equipment while wearing the show's iconic hazmat suits and sporting face masks on top of their heads (and hence without the benefit of their protection for eyes and lungs (S4E6). Two of the women were workers at the real-life laundry facility, a fact that the more privileged, above-the-line workers on the show seem condescendingly proud of in their DVD commentary:

> BRYAN CRANSTON ("WALTER WHITE"): These lovely ladies actually work there.... Two of them actually *do* work at this *real* industrial laundry that we shot at.
> GENNIFER HUTCHINSON (WRITER): And they don't speak English.
> CRANSTON: And they don't speak English.
> MICHAEL SLOVIS (CINEMATOGRAPHER): The fact they're all so short is just too tremendous.... They were game, and they were wonderful.
> JONATHAN BANKS ("MIKE EHRMANTRAUT"): Well, they *were* game, and they had such a good time, I would always see them afterwards and they were always laughing and smiling.
> VINCE GILLIGAN (CREATOR AND EXECUTIVE PRODUCER): Sweet ladies.

Breaking Bad uses warehouses in ways that indirectly suggest its reliance on low-wage racialized and gendered labor, visible in both its content and production. White's employment of laundry workers in the narrative is doubled by the show's employment of the women to film scenes at their everyday place of work, while the character's patronizing attitudes toward the women are partly matched by those of the more privileged people producing *Breaking Bad*. Here the show represents a kind of DEI TV in which structural racism parades as enlightened authenticity.[36] Finally, White's offer of $50 dollar bills ("*Presidente* Grant, very important man") recalls public subsidies for TV shows that create more low-wage jobs in New Mexico.

Settler Colonial TV

According to film historian Brian R. Jacobson, "the film studio embodies . . . cinema's broader worldmaking ambitions—an anthropocentric desire to control and simulate the nonhuman world."[37] Jennifer Faye makes a similar claim in her study of Buster Keaton's simulation of extreme weather events on elaborate sets, making his films "an early aesthetic paradigm of the Anthropocene."[38] Whereas extreme weather events were often created on soundstages, Keaton's efforts to create storms in physical locations reminds us of the expansiveness of the desires to control the natural world, or what Jacobson calls "the extrastudio logics whereby filmmakers "looked at 'natural' landscapes with studio eyes, reimagining what would come to be termed 'locations' as potential sets that could be mined and extracted to generate cinema's human-built virtual worlds and with them a humancentric conception of nonhuman nature."[39] The concept of the Anthropocene is limited, however, by its assumption of a universal humanity when what it in fact names is colonial humanism. In the present context, I would emphasize that the desire to control the natural world, including Indigenous land, is a defining feature of settler colonialism. A controlling, extractive, private property perspective on the world isn't characteristic of humanity as such but of settler colonial humanity contra Indigenous epistemologies.

As Vicki Mayer argues, "How the film economy dovetails and supports this upward redistribution of wealth speaks to the colonial tendencies within the film industry itself. Much like the tourism industry, Hollywood has always aspired to spatial domination." TV production is a "profit-driven hermeneutical enterprise," which includes "rationales and reflective analyses" justifying inequalities in screen industries.[40] As I show in subsequent chapters, TV's DEI narratives are just such cultural occupations, presenting diverse representations at odds with the reality of their production.

The examples of New Mexico and Georgia highlight how the designation of land as public is a means of expropriating Indigenous land and resources and a route to their privatization in the form of state subsidies to studios. Georgia's Blackhall/Shadowbox is a settler studio in that way. The sensitive forest land it acquired in the swap with the city of Atlanta historically belonged to Muscogee (Creek) Nation before they were removed in 1821. On November 27, 2021, Muscogee (Creek) Nation and Muscogee (Creek) descendants, including former Emory Professor Craig Womack and other migrants from Oklahoma, rallied in protest at the Intrenchment Creek Trailhead against Cop City and the studio expansion, giving speeches and performing a Stomp Dance. It was reportedly the first ever collective migration of Muscogee (Creek) Nation to their ancestral homeland. In April 2022, Womack joined University of Oklahoma Professor Laura Harjo and others for two days of events at Intrenchment Creek, including a second Stomp Dance.[41] The history of Muscogee (Creek) displacement is complemented by the exclusion of Indigenous people and stories in films and TV shows made in Georgia. In contrast with most productions, removal is indirectly referenced by the name of the cannibal settlement in *The Walking Dead*. As urban studies scholar Dan Immergluck reminds us, "in 1837, after the Muscogee (Creek) and Cherokee had been violently expelled from much of northern Georgia, a settlement known as Terminus developed where downtown Atlanta is today."[42]

Watchmen also obliquely references settler colonialism. The show's white terrorists are called the Seventh Kavalry (sic) in honor of Custer's genocidal unit. Viewers discover in one episode that white nationalist senator Joe Keene (James Wolk) has given the chief of police and Seventh Kavalry member Judd Crawford (Don Johnson) a painting titled *Feats of Comanche Horsemanship* as a sort of settler colonial trophy. Keene is compared to a cowboy, and Judd even wears a cowboy hat, suggesting the entanglements of white supremacy and settler colonialism. This analysis of cowboy Klansmen also helps contextualize the show's references to the Rogers and Hammerstein musical *Oklahoma!*, the title song of which is a kind of popular anthem of settler colonial Americana.[43] While the brief representation of settler colonialism is suggestive, it is displaced from Georgia, where *Watchmen* was filmed, to its Oklahoma setting, thus furthering the "forgetting" of Indigenous Georgia. Meanwhile, *Lovecraft Country*, which was partly filmed on Muscogee (Creek) land at Blackhall/Shadowbox, briefly gives voice to Indigenous history only to violently silence it. In episode 4, Leti, Tic, and Montrose discover a subterranean world under the statue of a white supremacist sorcerer. There they encounter a mummified group of Arawak people, killed by the sorcerer. When a two-spirit Arawak character named Yahima Maraokoti is reanimated, Montrose inexplicably slits their throat (S1E4). *Lovecraft Country* thus figures Indigenous worlds as part of a subterranean, ancient, and artifactual past with no place or voice in the present.

The state of New Mexico, including its studios, sits on Indigenous land. The area is home to 23 sovereign tribes, including 19 Pueblos, 3 Apache tribes, and the Navajo Nation. Native peoples make up almost 11% of the state's total population.[44] The theft of Indigenous land—and its possible return—continues to be a "live" issue in the present. In the inverse of the usual public-to-private land pipeline, the Obama administration placed in a trust a large parcel of land that had been appropriated from Isleta Pueblo and then sold it to the Pueblo. The location of the first film shot in New Mexico, *Indian Day School* (Edison Inc., 1898), the Pueblo borders Albuquerque Studios. For many years Isleta has purchased the naming rights to the Isleta Amphitheater, which is a part of the Mesa del Sol residential development, making the Pueblo a close neighbor of Netflix Studios.

The *Midnight, Texas* crew drove about 7 miles from Netflix to the Isleta Casino and Resort and used its lobby and glass elevator, with dramatic views of the surrounding Sangre de Cristo Mountains, to represent a hotel where the assassin Oliva Charity (Arielle Kebbel) kills one of her targets. In exterior shots, "Isleta Resort and Casino" is visible on a large sign, even though the scenes are set in El Paso. Also based at the nearby Netflix Studios, *Breaking Bad* filmed two scenes of a drug rehab center featuring the resort's striking spa building, designed to resemble Pueblo pottery, but which the show's producers misidentify as a kiva in DVD commentary (S2E13, S3E1). *Breaking Bad* returned to the resort for scenes of the White family in hiding at a local hotel, including Walter and Walt Jr. having a heart-to-heart talk by the pool. Walt and Skylar's hotel room was constructed on a soundstage at the studio but the set was modeled on those at the Isleta resort, with the addition of pottery and framed photos of the Navajo settlement To'hajiilee (S5E12) on the wall. About 30 miles west of Albuquerque, To'hajiilee is an important location for the show. Scenes in the first episode, of Walt and Jessie cooking meth, were filmed there, as were scenes of a fatal shootout between white supremacists and the DEA in the season 5 episode named "To'hajiilee." (See Figure 2.6.)

TV producers in New Mexico constantly acknowledge Indigenous peoples and lands in the present, if only because when filming at "To'hajiilee" and other Indigenous sites they must secure permission and pay fees. Such awareness does not, however, generally seep into the content of shows, which mostly avoid representing Indigenous contexts with any sensitivity or complexity. Even when they incorporate Indigenous characters and stories, shows shot in New Mexico generally sacrifice them for the benefit of narratives about white characters. Despite

Figure 2.6: DEA agents and white supremacist meth dealers face off at the Navajo settlement To'hajiilee in *Breaking Bad*.

its setting, *Breaking Bad* only includes a few Indigenous characters with names and dialogue. In season 1, a high school janitor named Hugo Archilleya (Pierre Barrera, Klamath/Lakota) is fired when he is blamed for stealing chemistry equipment taken by Walter White. In season 4, Gus Fring uses a box cutter to slit the throat of his henchman, Victor (Jeremiah Bitsui, Navajo/Omaha), and Walter and Jesse dissolve the body in acid. *Breaking Bad* thus renders Indigenous people disposable in narrative versions of settler colonial appropriation and elimination.

The combination of appropriation and elimination extends to the representation of Indigenous land. As Myles McNutt writes about *Breaking Bad*, "although To'hajiilee is central to the series' narrative as a setting," it is

> defined purely through its aesthetic and symbolic value to the story. It is a landscape that is given meaning through the storylines that unfold within it, but the location itself holds no agency over that story—in this case specifically, the Navajo Nation plays no significant role in the series' narrative, with the writers choosing not to engage with the cultural or political dimensions of those who own and govern the land in question. . . . Although the landscape is of significant value to the series, the place-making activities central to that community are less crucial.[45]

Breaking Bad and other shows shot in New Mexico often visually appropriate Indigenous landscapes while excluding from view the Indigenous lives connected to them.

Another way that studios reinforce settler colonialism is by constructing land as fungible. An oft-noted feature of contemporary film and TV production is the use of one place to represent another. This is true of many shows, but there are several examples from my case studies where the differences between settings and locations are particularly pronounced. *The Deputy* (Fox, 2020) used both Atlanta and Albuquerque to portray Los Angeles. Season 1 of *Good Girls* was set in Detroit but filmed in Atlanta. *In Plain Sight* employed Albuquerque locations for scenes set in Long Island, Boston, and Chicago. *Watchmen* used the Atlanta metro area to play Tulsa, Queens, and Saigon. Such differences between settings and locations are so common that they are the object of parody in *Get Shorty*, where a nineteenth century English costume drama is filmed on a dessert studio backlot (set in Nevada but filmed at the Netflix Studio in Albuquerque). New Mexico-based location manager Rebecca Puck Stair explains that the challenges of her job include "city-for-city doubling" and "landscape-for-landscape doubling."[46] The doubling of one city or landscape for another makes them interchangeable. As Noelle Griffis argues, "when one place doubles as another" it can "negate local distinctions" and contribute to the "erasure of local specificity."[47] Griffis describes a process of visual abstraction that complements the abstractions of capitalist exchange value whereby different commodities, including land, are rendered as abstract equivalents. Money, in other words, provides a way of representing value that negates distinctions between places and erases their specificity. It enables, for instance, the swap in Georgia of studio land for Indigenous land, represented as equal in value despite the local values tied to the land, values that are irreducible to money. Landscape-to-landscape doubling is the aesthetic corollary of a settler colonial perspective on land and resources as subject to exchange value in ways that are antagonistic to Indigenous epistemologies and relationships with land and the more-than-human world.

The difference between settler colonial doubling and Indigenous connections to land is suggested by the production history of *Reservation Dogs* (FX, 2021–2023). Focused on a group of Muscogee (Creek) youth and the adults around them, it is both set and filmed on Muscogee (Creek)

Nation land in Oklahoma, where co-creator and showrunner Sterlin Harjo [Seminole Nation of Oklahoma/Muscogee (Creek)] grew up. Executives at FX initially proposed filming the show in New Mexico, but Harjo successfully lobbied for the Oklahoma location. As he told the *Hollywood Reporter*, "I think that the authenticity that you feel in the show when you watch it comes from it being shot here. That is part of the tone. That specificity is the relationship to this place. And I couldn't fake it anywhere else."[48] In an interview with *Variety*, Harjo elaborated on some of the specificities of place that would have been lost if New Mexico had doubled for Oklahoma.

> Whenever you're making a show about Indigenous people, there's so much history of land displacement. We were brought here on the Trail of Tears from Alabama, Georgia, Mississippi and we were transplanted to Oklahoma, and then we rebuilt. There's so much tied into that. And there's so much history of character that's tied into that, and I had to shoot it here.[49]

By contrast, the Marvel program featuring an Indigenous woman superhero titled *Echo* (2024) was set in Oklahoma's Choctaw Nation but filmed in Georgia. Harjo's insistence that *Reservation Dogs* be filmed where it is set thus suggests a critical vantage point on location shooting as a settler colonial practice the makes different land fungible in ways that disappear their particularities and relationships to Indigenous peoples. Building on this chapter, the others suggest that state-subsidized TV production in poor places connects studio soundstages to settler colonial housing developments, police stations, prisons, and detention centers. The spatial networks of TV racial capitalism in place, I argue, represent the conditions of possibility for many contemporary shows and genres.

3

Crime TV

In 2015, the Albuquerque Police Department posted a photo to their Facebook page of 3 uniformed officers at a table eating a meal with *Breaking Bad* and *Better Call Saul* actor Jonathan Banks, who plays Mike Ehrmantraut, a fixer for brutal drug lord Gus Fring.[1] (See Figure 3.1.)

This was the year *Better Call Saul* first aired, so Banks was likely in town filming that show. The officer on the left in the back is Simon Drobik, an on-camera spokesperson for the department who also played police officers on the two programs as well as in the *Breaking Bad* spin-off film *El Camino*, where he was joined by other local police. He served as a technical advisor for all three works, helping the makers of

Figure 3.1: Jonathan Banks (back, right), who plays Mike Ehrmantraut on *Breaking Bad* and *Better Call Saul*, with members of the Albuquerque Police Department.

El Camino borrow a tactical vehicle from the Albuquerque Police Department. Officer Drobik was also the technical advisor for other programs I discuss, such as the crime drama *Briarpatch* and the show about gentrification, *The Curse* (chapter 7). Finally, he has worked as a stunt person for many films and TV shows shot in New Mexico, including a few shows discussed in this chapter—*Big Sky*, *The Cleaning Lady*, and *Deputy*.[2] Page and Ouellette argue that TV programs filmed in prisons infuse "the carceral state with cultural enterprise, social currency, and even glamour," and I would add that much the same is true for police who work in TV and become "micro celebrities."[3] In 2020, however, Drobik was forced to retire from the APD when it was revealed that he had fraudulently collected tens of thousands of dollars in overtime for police work he didn't do while he was instead working for Hollywood. He was routinely among the highest earners in the city, and ranked first among all city employees in 2018, making $192,973.[4] Drobik's appearance in the photo seems appropriate since he siphoned public funds for his own enrichment just as the show's producers appropriated state resources in the form of subsidies.

The grinning officer on the right in the photo, in front of Banks, is Jonathan Franco, who, at the time of the picture, had been accused of using excessive force. When a car thief stopped running, sat on the ground, and raised his hands, Franco kneed him in the head, a fact he left out of his report. In 2022, the city settled a lawsuit from the victim for $42,500.[5] In 2018, Franco was once again accused of excessive force when he and his partner kneed someone they were arresting and fractured his arm.[6] Then in 2018, Franco violated department policies regarding suicide watches by failing to check on an inmate every 30 minutes and the inmate hung himself in his cell. When he discovered the body, Franco also failed to check for signs of life or to provide any aide. His report on the incident included false information. Although the command staff recommended his termination, the police chief instead suspended Franco for 360 hours, which he was allowed to serve by taking a day or two of unpaid leave each week.[7]

Officer Franco is not one bad apple but representative of the Albuquerque Police Department's culture of violence and murder at the time. That was effectively the conclusion of a 2014 Department of Justice investigation. One legacy of *Breaking Bad* is that the show has lent its name to the titles of articles about the department scandal, such as the *Rolling Stone* investigation by Nick Pinto titled "When Cops Break Bad: Inside a Police Force Gone Wild." As Pinto wrote in 2015, "In the past five years, the police department of Albuquerque, a city of just 550,000, has managed to kill 28 people—a per-capita kill rate nearly double that of the Chicago police and eight times that of the NYPD. Until now, not one of the officers in those 28 killings had been charged with any crime." The reporter's interview with an anonymous former APD officer is particularly chilling: "'The focus was no longer on the mission as I understood it. . . . It was more about shooting people—as much as you could do so legally. The new culture was: 'Anybody you could shoot.' The officer had always railed against representations of the police as violent thugs, but 'to come to the end of my career and see that it was true—it totally messed me up.'"[8] TV shows that film in New Mexico often implicitly support police cultures of violence by subsidizing the police and representing them in a positive light.

State incentives to attract film and TV production to precarious places, in other words, are part of a police and prison televisual complex. TV racial capitalism in place is carceral, both in representation and in the practical alliances between Hollywood and local police and prisons. Crime dramas are among the most popular TV genres, and, as I will argue, many made in Georgia and New Mexico burnish the reputation of the police, representing them as champions of diversity. While they sometimes depict bad cops, on balance TV police look good, particularly when contrasted with white nationalist MAGA antagonists, or with Mexican narcos. When not spotlighting DEI police, in other words, TV shows foreground forms of white nationalism and racialized criminality that deflect attention from Hollywood's appropriation of public resources diverted from poor Black, Latinx and Indigenous people.

TV, the Police, and Prison: Partners in Crime

Studios and criminal justice institutions occupy related geographic and political economic spaces in Georgia and New Mexico. The geographic proximity of police, prisons, and studios is a result of their adjacency, or what they have in common, like adjacent rooms that share a wall, or adjacent angles that share common sides. In Georgia and New Mexico, carceral institutions and studios are physically near each other because they are often sited on state or federally owned land as in the first case, or on land that corporations purchase from states at substantial discounts in the second. Carceral institutions and film and TV studios also depend on the state construction and maintenance of roads and electrical power, water, and sewer systems. With state subsidies for land, studio infrastructure and production costs, film and TV studios share material spaces and funding models with police facilities, prisons, jails, and detention centers. The critique of prisons that they divert funds that could otherwise fund education applies to TV production as well, albeit on a smaller yet still significant scale. Given their shared interests, Hollywood and the police are practical allies. By advertising the supposed benefits of state subsidies, TV studios also advertise state power, one face of which is the criminal justice system.

The public land Netflix leased in 2020 with significant state and city subsidies for its Albuquerque studio expansion is a brief walk from Homeland Security and DEA offices, two large, warehouse-like structures resembling studios in exterior appearance. Partly filmed at the studio, *Breaking Bad* employed their DEA agent neighbors as technical advisors to help prevent the show from teaching viewers how to cook meth, and to design illegal meth labs, including the iconic RV lab and the lab beneath the Lavandería Brilliante (chapter 2).[9] Netflix Albuquerque Studios have also hosted *Stranger Things, Daybreak, Preacher*, and numerous other programs. Meanwhile, Santa Fe Studios (*Longmire, Roswell, New Mexico, Succession*) is minutes from the New Mexico Correction Department Central Administration, the Santa Fe County Adult Detention Facility, the Penitentiary of New Mexico, and the New Mexico Corrections Department. The state-subsidized "Santa Fe Studios Rd." connects the studio to the main gate of the Corrections Department and to the parallel "Penitentiary Rd.," leading to the penitentiary and to "Camino Justicia," leading to the county's adult detention facility.

In Georgia, both Shadowbox/Blackhall Studios (*Lovecraft Country*) and the police training center called Cop City depend upon provisions of adjacent public lands, the latter in the form of 85 acres in an unincorporated part of Atlanta. The training center will be built over the decaying remains of the Old Atlanta Prison Farm. Built in 1918 on the site of a former plantation, the farm required inmates to work long days growing crops and butchering hogs until its closure in 1995. In the intervening years, the decaying prison was used as a filming location.[10] The same activists who oppose the studio expansion also oppose the training center, since they hoped the prison property could be turned into public greenspace. As Jacqueline Echols of the South River Watershed Alliance told an interviewer,

> These two developments (Cop City and Shadowbox/Blackhall Studios) almost occurred in tandem. They did not evolve or develop gradually; they happened over a brief ten-year period. . . . The chain of decisions leading from the county's initial disinvestment to the emergence of Cop City and Ryan Millsap's Blackhall Studios provides a vivid example of how vulnerable and resource-deprived communities can be quickly consumed by inequity. These communities are the least able to fight off such acts of violence.[11]

In addition to sharing endangered river and forest spaces, the studio and police training center will both employ constructed sets, the latter in the form of mock inner-city neighborhoods for police training.[12]

Mediating the relationship between TV production and carceral institutions, studios construct police department and prison sets and incorporate police characters. Season 3 of *The Walking Dead*, for instance, juxtaposes two settings, the old industrial warehouses that are part of the antagonist Woodbury community and the "West Georgia Correctional Facility" (a backlot set), where the show's protagonists seek refuge. Tyler Perry Studios outside Atlanta includes a permanent county jail set and a prison yard set.[13] As for shows shot in New Mexico, crime dramas such as *Better Call Saul*, *Breaking Bad*, *The Cleaning Lady*, *The Deputy*, *In Plain Sight*, *Killer Women*, and others, regularly employ jail sets. Finally, film ranches in New Mexico such as the Bonanza Creek Ranch and the J.W. Eaves Movie Ranch and others feature old west jails sets.

TV producers also shoot scenes in real prisons. New Mexico state law grants film and TV production companies access to publicly owned buildings in exchange for modest security fees, including the now closed "Old Main" New Mexico State Penitentiary, which is also a police training facility.[14] The New Mexico Corrections Department website includes information about filming locations at the Old Main, which are depicted

Figure 3.2: The watchtower of Santa Fe, New Mexico's Old Main Penitentiary in *MacGruber*.

Figure 3.3: The Old Main Penitentiary in *The Lost Room*.

in a slide show. The website also advertises the availability of corrections officers, for an hourly fee, to act in films and TV shows.[15] The penitentiary was used to represent a mysterious portal to another dimension where a white police offer searches for his daughter in *The Lost Room*; a jail where the title character of *Terminator: The Sarah Connor Chronicles* is incarcerated; a military research facility and detention center for aliens in *Roswell, New Mexico*; a prison where the title character of the comedy *MacGruber* is incarcerated; and a prison for sex offenders in *Perpetual Grace, Ltd* (See Figure 3.2.)

A short drive from the Netflix Albuquerque Studios, *Breaking Bad* used the Metropolitan Detention Center to shoot a montage of murders set in three different prisons.

Meanwhile, the new State Penitentiary that replaced the Old Main played the fictional federal prison "ADX Montrose" where Kim Wexler visits Jimmy McGill in the series finale of *Better Call Saul*, and the Montana prison where Walter Barnes is unjustly incarcerated for murder in *Big Sky* (S3E10).

Finally, playing itself, the Penitentiary was featured in an episode of the witness protection program *In Plain Sight* (S3E6). Such shows seemingly align themselves with the perspective of panoptic prison towers, as in *Breaking Bad* and *Better Call Saul*, which include high angle shots of prisoners and penitentiary yards that mimic prison surveillance. Complementing Hollywood's dependence on low-wage racialized labor more generally, many New Mexico shows employ Chicanx and Indigenous extras to play prisoners in orange uniforms.

Home to Eagle Rock Studios, Georgia's Gwinnett County boasts several carceral filming locations. The local film office advertises the Lawrenceville Jail, which once held slaves, as well as the Snellville City Jail, the Gwinnett County Police Training Center (which the new center next to Shadowbox/Blackhall Studios will replace), and the Gwinnett County

Figure 3.4: Albuquerque Municipal Detention Center in *Breaking Bad*.

Figure 3.5: Kim Wexler leaves the Penitentiary of New Mexico in Santa Fe after visiting Saul Goodman in *Better Call Saul*. Such high angle shots of prisoners and penitentiary yards mimic prison surveillance.

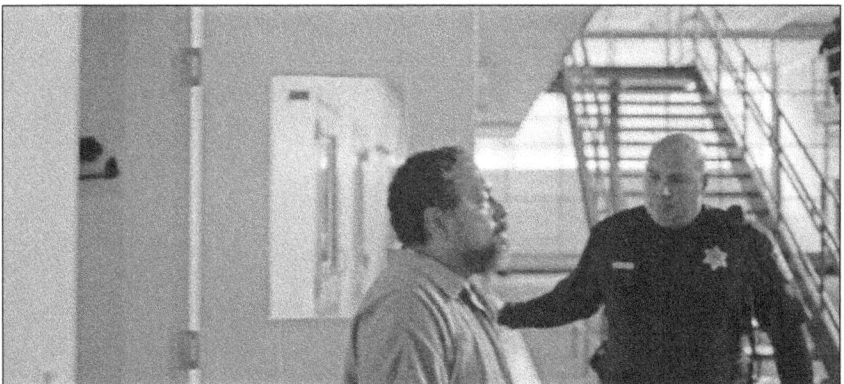

Figure 3.6: A prisoner on his way to the visitation room in the Gwinnett County Detention Center, Lawrenceville, Georgia, on the show *Ozark*.

Detention Center, in Lawrenceville.[16] In addition to films, the TV programs *Greenleaf* (S3E7–8), *Rectify* (S1E1), *The Outsider* (S1E2), *Insatiable* (S2E4), *The Staircase*, and *Will Trent* (S2E2) have employed the Gwinnett County Detention Center to represent prisons where white characters are incarcerated, while season 4 of *Ozark* used it to play the prison holding Mexican drug lord Omar Navarro. (See Figure 3.6.)

In the words of a regional TV news report, "Local jail reaps benefits of Georgia's booming film industry." The warden featured in the report explains that the fees paid by production companies are donated to charity, and that film companies employ off-duty guards for security.[17] On the one hand, TV producers benefit from low rates for locations that would be expensive to reproduce on studio soundstages or backlots. On the other hand, the prison benefits from opportunities for positive public relations via connections to charity and Hollywood while partly offloading the cost of reproducing prison guard labor by supporting their employee's side hustles working for the studios. Similarly, *The Walking Dead* helped fund the Senoia Police Department. According to Chief J. E. Edens, in one year alone the show paid over $17,000 in fees for use of police vehicles, which the small-town department used to fund a K-9 unit.[18] The department's website effectively advertises the police by featuring a large background photo of an iconic portion of the town that was used in *The Walking Dead* to play the survivor settlement Woodbury. And most recently, the Netflix drama *A Man in Full* (2024) advertised Atlanta's infamously deadly Fulton County Jail by filming there.

As the example of Officer Drobik and his work on *Breaking Bad* and other TV shows suggest, many programs also promote the police by employing them as advisors and actors. In their critical account of the Georgia-filmed *Watchmen*, for example, Jessica Hatrick and Olivia González note that one of the show's writers was a former Chicago police officer, and the program employed a local law enforcement officer as a police coordinator and featured a celebrated local K-9 and their police sergeant handler. Police employment is complemented, Hatrick and González argue, by *Watchmen*'s content, which they read as pro-police propaganda or "copaganda." Despite its

> critiques of white supremacy and the historical ties between policing and white supremacist organizations, *Watchmen* ultimately reproduces copaganda conventions and fails to show policing as a structurally white supremacist organization. Through portraying central police characters who commit violence as heroes, uplifting the main cop character as an eventually almighty arbiter of justice, and portraying the Seventh Kavalry's power as limited, *Watchmen* tells audiences that police violence is

necessary, noble, and justifiable; forwards a notion of justice rooted in state punitivity; and depicts white supremacy as fringe and exceptional.[19]

Together these examples suggest the symbiotic relationship between studio warehouses and the warehousing of people.

The study *Normalizing Injustice: The Dangerous Misrepresentations that Define Television's Scripted Crime Genre* by media activist group Color of Change demonstrates that TV programs about the police generally support the perspective of criminal justice professionals over those of critics and reformers, let alone abolitionists, largely excluding representations of racial inequality in terms of profiling, bail, and sentencing.[20] In the present contexts, I would add that with few exceptions, shows shot in real carceral institutions foreground white protagonists in locations where large numbers of people of color are disproportionately imprisoned.[21] This means that above- and below-the-line TV production workers pass through razor wire and chain-link fences, set up scenes in prison yards, visiting rooms, and cafeterias, and then help tell stories about white people facing criminal justice or injustice. Confronting real prisons and then constructing misleading fantasies represents a stunning form of disavowal—I know what prison is like and yet all the same I'll act like I don't. TV shows filmed in real prisons thus obscure racial inequality in general, but particularly in the poor places where they are produced. Such shows disavow the historic roots of criminal justice in Indigenous and Black slavery, histories of prison labor, immigrant detention, and ongoing forms of racialized violence, police shootings, and prisoner deaths. Since 2019, eight inmates at the Gwinnett County Detention Center have died while in custody. Fifteen inmates have died since 2020 at the Metropolitan Detention Center used in *Breaking Bad*.[22]

DEI Police

As depicted on TV, police departments are among the most diverse, equitable, and inclusive institutions in the United States. Set and filmed in and around Albuquerque, the blonde, white woman protagonist of *In Plain Sight* (2008–12) is a tough but fair US Marshal working in the Witness Protection Program named Mary Shannon. Similarly, as described on Imdb.com, *Killer Women* (ABC, 2014), set in Texas but filmed in New

Mexico, "centers on Molly Parker, one of only two women in the notoriously male Texas Rangers, a ballsy, beautiful badass who knows how to get to the truth and isn't afraid to ruffle a few feathers on her way there." Although the Texas Rangers have a history of sexism and anti-Mexican violence, including lynching, Parker is a new kind of DEI Ranger, with a lovably gruff Latino chief and a Latina best friend. The blonde Ranger even moonlights playing trumpet for a local Tejano band. Set in Elena, Montana but filmed in New Mexico, *Big Sky* (2022–2023) also features a blonde police officer named Jenny Hoyt (Katheryn Winick) and a Black private detective named Cassie Dewell (Kylie Bunbury). All three shows focus on conventionally attractive white women police officers, thus combining liberal white feminism with eye candy.

Police programs thus demonstrate their commitment to DEI representations in terms of sex and gender identities. *Big Sky*, for example, includes the first openly transgender series regular on ABC, private detective Jerrie Kennedy (Jessie James Keitel), who in one episode confronts the memory of her trans-phobic parents. Created by Will Beal, a former LAPD detective, the Fox show about the LAPD titled *Deputy* (filmed in both Georgia *and* New Mexico) also features a trans character, Deputy Brianna Bishop. The character's narrative arc suggests that, in contrast with Bishop's personal life, the LAPD is tolerant and welcoming of diversity. Bishop's femme-identified Black and Asian girlfriend Genevieve (Karrueche Tran) ultimately scorns them when they come out as non-binary, but Bishop's co-workers embrace them, especially the cowboy hat and boot-wearing white Deputy Bill Hollister (Stephen Dorff, S1E9). (See Figure 3.7.)

The rest of the show's characters are "diverse," including Hollister's Latinx wife, Dr. Paula Reyes (Yara Martinez), and Black deputies Joseph Harris (Shane Paul McGhie) and Charlie Minnick (Danielle Moné Truitt).

The show *Walker: Independence* (2022), a New Mexican-made prequel to *Walker, Texas Ranger* set in a late nineteenth century frontier town, includes a similarly diverse cast of characters: the original show's ancestor (a blonde white woman), a heroic lesbian Pinkerton, a Chinese laundryman, an Apache warrior, and a Black deputy sheriff. Over the course of the show, these characters become allies in a fight against rapacious railroad executives. With its DEI mise-en-scène, *Walker: Independence*

Figure 3.7: Trans Deputy Brianna Bishop fists bumps a fellow officer on *Deputy*.

problematically redeems the Pinkertons, private detectives infamous for their anti-worker violence and, implicitly, the Mexican-murdering Texas Rangers. The Georgia-set and -filmed show *Will Trent* (2023–) adds neurodiversity to the DEI mix. The title character (played by Ramón Rodríguez) is an officer in the Georgia Bureau of Investigation and a former ward of Atlanta's overwhelmed foster care system. Trent is dyslexic and, while hinted at earlier, in the final episode of season 1 it's revealed he's Puerto Rican. He is joined on the force by African American officer Faith Mitchell (Iantha Richardson); Faith's mother and former officer Evelyn Mitchell (Lisa Gay Hamilton); Will's friend from foster care and sometimes-romantic partner and recovering drug addict, white officer Angie Polaski; and his Black captain Amanda Wagner (Sonja Sohn). In season 2, this diverse group of police come to the aid of a drag club threatened by a homophobic gunman (S2E8).

In addition to the diversity of cast and characters, many shows suggest that police are agents of anti-racism, opposed to white nationalism and the MAGA movement. *Deputy*, for example, begins with officer Hollister being disciplined for tipping off migrants to an ICE raid. Later, when he becomes Sheriff, he stops the department from aiding ICE.[23] Over the course of three seasons, the women officers and detectives in *Big Sky* repeatedly face off against patriarchal white men who violently prey on women, including a serial killer named Buck Barnes (Rex Linn), who murders women and cuts out their hearts (S3). With his western wear, trucker's hat, and white goatee, Buck looks like a TV version of a MAGA supporter, implicitly suggesting the liberal perspectives the show presumes. In one episode of *Will Trent* inspired by real world struggles over Confederate monuments, Trent discovers a plan by a white nationalists called the "Confederate Front" to attack people protesting a Confederate memorial. As their leader tells a group of white supremacist terrorists, "Liberal snow-flake politicians want to take down a Confederate statue? Not today boys, so grab a gun and get to know it" (S1E7). But the show's trio of a Latino agent and two Black women agents, Mitchel, and Wagner, join forces to arrest the neo-Nazis.

Although most programs in my sample focus on state policing, at least one Georgia-made series represents a DEI world of private policing, the dark comedy *Teenage Bounty Hunters* (2020). The title characters are wealthy, white twin sisters who attend a Christian private school, Sterling and Blair Wesley (Maddie Phillips and Anjelica Bette Fellini). In the first episode they convince a Black bounty hunter named Bowser Jenkins (Kadeem Hardison) to hire them because they can help him navigate their parents' racist country club in pursuit of a fugitive. The teens track the fugitive to the club's "bourbon room," decorated with a Confederate flag. In subsequent episodes, the sisters' wisecracks deflate white nationalism. When they learn that there are 134 Confederate statues in Atlanta, Sterling exclaims "Dang! They know the South lost right?"; and when they visit the Confederate Historical Society for research, Blair says "Looks like a jerk-off booth for *The Dukes of Hazzard*" (a show that was filmed in Georgia). Created by Kathleen Jordan, daughter of Hamilton Jordan, chief of staff for President Jimmy Carter, *Teenage Bounty Hunters* is inspired by her liberal upbringing in the wealthy, conservative Bucktown area of Atlanta. In interviews, Jordan emphasizes the show's sex-positive depictions of the teens, and it is also humorously aligned with liberal anti-racist representational politics.[24]

Police programs combine DEI representations and racist images of Mexican drug traffickers as two sides of the same coin. In the first episode of *Killer Women*, Ranger Parker illegally crosses the border into Mexico with a DEA agent to rescue a Mexican grandmother and her granddaughter from a Mexican cartel. Also working with the DEA,

Agent Will Trent goes undercover as "Bill Black," an enforcer in a Latinx drug gang, to identify a mysterious Latinx "drug lord" only known as "CK." He ultimately discovers that the drug lord is a Latina named Rosa, who claims to be studying to be a nurse but is secretly a killer. In the process of stopping her, Trent saves a trans woman from being murdered by the drug lord. Meanwhile, *Deputy*'s depiction of Mexican immigrant criminal violence complements news coverage on Fox, the show's network. Episodes focus on a Mexican drug cartel and human trafficking ring that uses gangs and a Mexican bank to launder money in the form of remittances. Indeed, menacing Mexican gangsters with face tattoos are the program's central villains. *Deputy*'s seemingly progressive, liberal perspective is combined with plots echoing Donald Trump's infamous claim when he announced his run for the Presidency in 2015: "When Mexico sends its people, they're not sending their best. They're not sending you. They're not sending you. They're sending people that have lots of problems, and they're bringing those problems with us. They're bringing drugs. They're bringing crime. They're rapists. And some, I assume, are good people." As we see here and in the subsequent section, Trump's statement could serve as a synopsis for many police shows. To be sure, the combination of DEI police and Mexican narcos is ubiquitous in US film and TV, but it assumes particular significance in New Mexico and Georgia, given that TV producers there have sustaining, mutually beneficial relationships with prisons and police. DEI elements distract from the racism in representations of Mexican immigrant gangsters and cartels so that such depictions are naturalized as foils defining law enforcement heroism.

In his study of the Los Angeles Police Department's DEI programs, Dylan Rodríguez concludes that "the LAPD's multiculturalist refurbishing sustains rather than interrupts the force of racist state violence."

> Consider the Los Angeles Police Department's conspicuous efforts to augment and publicly signify the demographic diversity of its officers in the immediate aftermath of the late 1990s Rampart Division scandal, a massive police corruption case that implicated over 70 officers, led to the reversal of more than 100 false criminal convictions, and resulted in close to 150 civil lawsuits against the city.... (These included) sponsorship of

the 2006 Gay Games Sports and Cultural Festival (a national initiative to increase the numbers of out gay cadets), co-sponsorship of the Tenth Annual International Criminal Justice Diversity Symposium (2006), and the ongoing "Join LAPD" recruitment imitative.

Of note in the present context is Rodríguez's analysis of the "Join LAPD" program as a form of police visual culture that legitimates ongoing racialized violence by giving it a DEI face. As he writes, "the pedantic visual solicitation of a diversified (cisgender, women, Latinx, Black, LGBTQ, and so on) cadet profile pervades the advertising and social media platforms of the Join LAPD campaign." The initiative also included a widely visited Facebook page, as well as freeway billboards, a particularly important form of visibility in car-centered Southern California. The implication is that "the visual apparatus of police multiculturalism" is not just about recruiting new kinds of police but also about advertising policing among liberal people in Los Angles, where so many film and TV producers live. "Between the subtle photographic display of the rainbow gay pride flag on the windshield of an LAPD squad car and an Instagram commemoration of 2019 International Women's Day, Join LAPD provides a living archival index of racial state multiculturalism as a logic of martial rearticulation that expands as it reforms the logics and protocols of" racialized policing. To remain "politically and institutionally viable" in the wake of the Rampart scandal and other instances of police violence and corruption, "white supremacist policing must undergo substantive reform" by incorporating nonnormative bodies and projecting "diverse" images of the police to the public. Indeed, as Rodríguez notes, diversity recruitment and hiring practices have not changed the LAPD's disproportionate targeting of Black and Latinx people, and Los Angeles remains home to the largest jail system in the world, "a carceral toll overwhelmingly borne by Black and Latinx people (including youth)." As I have argued elsewhere, during the Reagan-era war on drugs and after, the federal government worked to turn Hollywood into an ally by encouraging anti-drug, anti-trafficking plots in film and TV shows while also providing access to expensive military and police equipment.[25] More recently, Hollywood has reproduced such police partnerships in New Mexico and Georgia. In addition to the practical and material symbiosis between TV and the police, with its anti-narco representations Hollywood also brings the ideological infrastructure that supports racialized policing, criminalization, and carceral violence to the two states. This is one facet of the reinvention of the police as a culture industry in league with TV producers to promote Mexican criminalization.

The use of TV to solicit empathy for the police is allegorized in an episode of *In Plain Sight* titled "Training Video" (S2E12). When the department hires a film director to update its old training video, agent Shannon reluctantly agrees to work as a technical advisor, recalling actual police advisors like APD's officer Drobik. The director gives her the script and asks for notes, and so she visits the set to observe. Witness Protection agents cannot tell loved ones about their job, and in the first version of the video, when a woman agent confesses to her male supervisor that the silence is straining her marriage, the supervisor replies that she should just focus on the good parts of the job. Shannon presses the director to do better, and in the final version of the video, the supervisor admits that he too struggles to balance work and his private life while advising her to explain to her husband that the silence is hard on her too. When she follows his advice, her marriage improves since both feel their struggles are appreciated. The video within the show seems to represent the aspiration of many TV producers to "humanize" the police and make them "relatable."

The Work of Police Washing: *Devious Maids* and *The Cleaning Lady*

Carceral systems are like Hollywood's accomplices since prisons and the police control bodies and populations within the contexts of the low-wage, racialized, and gendered labor that benefits TV production in poor places. TV maids of color clean up the crimes of the wealthy just as actual woman of color cleaners service local economies based in TV production and their police and prison partners. *Devious Maids* (2013–2016), where the maids of the title clean crime scenes and cover for the wealthy, symbolically resolves conflicts or contradictions between workers and employers by scapegoating Mexican narcos. The show is a dark comedy about a group of Latinx maids who work for the dysfunctional and scandal-prone rich, including by cleaning murder scenes. Set in Beverly Hills, the pilot was filmed in Los Angeles but when ABC declined to pick it up, the show was purchased by the Lifetime network and production shifted to Atlanta, first at EUE/Screen Gems but then for three seasons at Eagle Rock Studios. As the producers have noted, the Georgia incentives made it more "economically feasible" to film *Devious Maids* on a smaller, cable budget.[26] In content, *Devious Maids* mystifies the labor relationship between the maids and their employees. Rosie marries her employer, TV soap opera star Spence; Marisol and her boss Taylor become best friends; Zoila and her employer Genevieve are like sisters. In her essay titled "Television for All Women? Watching Lifetime's *Devious Maids*," Jillian Báez argues that in an effort to attract both white women and Latina viewers, the show adopted a postfeminist, postracial perspective that "de-emphasizes gender and racial inequality within the narrative but appears inclusive through a racially diverse cast."[27] This dual audience address helps explain the friendships on the show between employers and maids, but I would add that such friendships are fortified by a shared fear of Mexican narcos. Despite their differences, the show suggests, the rich and their maids are more like each other than the terrifying Goviota cartel, which threatens to murder the young daughter of Marisol's employers. *Devious Maids* thus downplays the racism faced by Latinx maids while reproducing racist representations of Mexican narcos. Once again Mexican cartels represent the naturalized nadir of evil, which makes Hollywood and the police appear benevolent by contrast. The show exploits low-wage service labor materially, in its production, and in its content, where depictions of diverse maids implicitly sanitize the image of the police relative to "dirty" Mexican criminals. Latina producers and Latina actors helped make a show that profited from raced and gendered low-wage labor in Atlanta while representing the work of Latina maids as exploitation-free, and while figuring Mexican narcos as the sort of the threat justifying police power to keep poor and working-class Black and Latinx people under control. *Devious Maids* thus exemplifies Latinx racial capitalism in place, in which privileged Latinx creatives profit at the expense of underprivileged Black and Latinx communities who are disadvantaged both by the diversion of public resources to Hollywood and the expanded support for local police, both practically and in terms of representation, that defines the police and prison televisual complex.

Figure 3.8: On *The Cleaning Lady*, Thony is placed in ICE detention.

Most of the shows in my sample seem to direct our attention inadvertently, almost accidently, to the mutually beneficial relationships between TV production and the criminal justice system that buttresses racial capitalism in place. *The Cleaning Lady*, however, more explicitly and seemingly consciously connects the dots between labor, TV production, and prison. In Las Vegas, the Cambodian migrant from the Philippines Thony De La Rosa works cleaning hotels and warehouses, but when threatened with deportation she is forced to clean crime scenes for an arms dealer. In one episode, Thony, her sister-in-law, and other undocumented women workers are arrested by ICE and detained in chain-link cages in a large warehouse. (See Figure 3.8.)

According to the show's creator and showrunner, Miranda Kwok, the episode was inspired by an August 2019 raid of food processing plants in Mississippi resulting in the separation of families and the detention of 680 workers. As if orchestrated to produce maximal terror, the raid was scheduled for the first day of school.[28] The show's ICE detention sets were based on images by photojournalist John Moore of children and adults held at the Ursula Detention Center in McAllen, Texas, complete with the ubiquitous silver mylar blankets.

Itself a retrofitted warehouse, the Ursula Detention Center was recreated on a studio soundstage for *The Cleaning Lady*, which visually quotes Moore's photos, including the famous image of a detained migrant child watching a television attached to a chain-link wall. (See Figure 3.9.) Such images suggest that TV is part of the infrastructure of detention, perhaps in recognition of the fact that TV production in New Mexico depends on undocumented workers whose low wages are a feature of their deportability. In contrast with the skilled workers directly employed in TV, the undocumented women workers who clean up after them are non-unionized and vulnerable to threats of deportation if they make any demands, a situation repeatedly dramatized in *The Cleaning Lady*.

With their depiction of Thony, Kwok and Yung hoped to undermine both the model minority myth and depictions of immigrants as criminals by focusing on the precarious positions of undocumented women workers.[29] *The Cleaning Lady* thus makes visible what the other shows in this chapter obscure, reminding us that the symbiosis between TV and the police is also a form of labor control. Racial capitalism in place is carceral, the show suggests, not only because the police protect Hollywood's properties but also because they help contain the low-wage workers who, largely excluded from jobs in the TV production, nonetheless make it possible.

Figure 3.9: ICE-detained children watching TV, *The Cleaning Lady*.

4

Horror TV

The spread of COVID-19 in Georgia transformed the final season of *The Walking Dead* (AMC 2010–2022), the popular program about a zombie virus. The showrunners faced a dilemma: Georgia state subsidies made shooting there attractive, but it was also potentially dangerous for large numbers of actors to come into close contact as part of zombie hordes. The virus thus changed the narrative of the show's final season, forcing producers to focus more on individual character arcs and stories requiring fewer extras so that *The Walking Dead* could continue to benefit from Georgia's lucrative subsidies.[1] COVID-19's effect on the program's narrative structure opens a window on the workings of TV racial capitalism in Georgia. Material inequities there ensured that the virus disproportionately affected the state's Black residents.[2] Georgia has the second lowest level of state spending for social welfare in the United States, and the percentage of Black people in Georgia living below the poverty line is significantly higher than the national average. As a result, Black people in the state have been more vulnerable to the virus because they live in crowded housing, perform essential yet low-wage and dangerous work, have limited access to health care, and suffer from the negative health consequences of racism.[3] The COVID-19 pandemic is a unique, once-in-a-lifetime crisis with unprecedented consequences for life and labor, yet the exceptional global disaster has drawn into relief the local, quotidian conditions that make people of color especially vulnerable to illness and premature death. The dangers of shooting *The Walking Dead* during the pandemic are novel, but they remind us that Georgia is Black and poor even in "normal" times, and those conditions are part of what enables the state to attract Hollywood media makers.[4]

Horror shows like *The Walking Dead* indirectly reference the terrors of poverty, inequality, and asset stripping, or what Clyde Woods referred to as all the "tricks and traps" poor Black and Latinx people face in the neo-plantation economy of Louisiana, as well as, I would add, in the distinct but related contexts of Georgia and New Mexico.[5] *Atlanta* and *Lovecraft Country* in particular invite investigation into the violence (both slow and fast) that accompanies location shooting and the construction of film studios, violence which the TV shows themselves often obscure, distort, or even seemingly endorse. Hence while hinted at, by and large the horror shows in this chapter disavow the material inequalities on which they depend. A trio of programs about serial killers—*The Unsettling*, an episode of *Atlanta*, and *Swarm*—depict the horrors of foster care as a legitimating other for TV production, which is implicitly represented as a better use of state funds. The shows effectively suggest that TV production makes a greater contribution to diversity than social spending. Meanwhile, a group of supernatural horror shows with diverse casts and anti-racist narratives—*Midnight, Texas, Sleepy Hollow, Preacher,* and *Lovecraft Country*—nonetheless aestheticize and rationalize forms of disposability characteristic of TV racial capitalism in place: the disposability of Black women workers, communities of color, and the environment. Ultimately, the horror programs analyzed here use genre conventions to construct a DEI gaze that disavows local conditions of production, making it more difficult see and critically reflect on TV's role in reproducing racial capitalism in place.

DEI Serial Killer Horror

The Unsettling (Netflix, 2019), *Atlanta* (FX, 2016–2022), and *Swarm* (Amazon, 2023) employ the conventions of horror to represent the violence of state foster care systems. To be sure, real-world foster care can be problematic in its overlap with carceral institutions, but the three programs depict foster care in extreme, horrific ways. Foster care, I argue, symbolically stands in for the kinds of social welfare from which TV and film production diverts resources via state subsidies. Horrifying representations of foster care implicitly justify Hollywood's upward redistribution of wealth as a better use for public monies.

Filmed in Santa Fe and Lamy, New Mexico, *The Unsettling* is about a multicultural group of teens in foster care. Their foster parents, Fia and Jason Warner, are in an apocalyptic white Christian death cult called "New Purity" led by Fia and a sinister character called Elder Isaac, who totes a rifle and sports a wispy Jesus beard. Together they plan to burn the foster children to death as a sacrifice and "heal the land," a plan telegraphed in episode 5 when Fia whips herself in a garage filled with gasoline cans. Fia ultimately murders her husband with a farming tool when he refuses to help, and unsuccessfully attempts to burn the children in a barn at gunpoint. The foster mother's scheme depends on the disposability of the vulnerable, multi-cultural teens. As Becca tells the others, "We were all chosen for a reason—we'd be invisible if we disappeared." Or as Fia asserts, the foster kids are "just tinder for the fire." *The Unsettling* seems partly inspired by contemporary political extremism, with New Purity's commitment to regeneration through violence suggesting comparison to some forms of MAGA fundamentalism. Indeed, the MAGA movement's denial of the seriousness of the COVID-19 pandemic and resistance, sometimes violent, to safety measures often inspired commentators to call it a "death cult."[6] *The Unsettling* refigures that contemporary far-right political context in a narrative about a cult of white Christian serial killers.

The program's critical engagement with white Christian nationalism is subtly suggested by the choice of setting for the foster family, a compound called Crow Ridge Ranch and a large "territorial revival style" New Mexican home. That style of architecture emerged in the 1930s as a revival of elements from buildings produced by white US Americans in New Mexico when it was a territory. As a revival style, it represents nostalgia for an era of white colonial rule over Mexican and Indigenous people, a time before even the semblance of democratic representation in the New Mexican territory. Examples of this architectural style in Northern New Mexico include the New Mexico Capital Complex and other government buildings; many Christian churches; and homes like the one in *The Unsettling*. As in other horror films, with its locked basement, spectral mirrors, and menacing observers peering out of attic windows, the house is like a character in the show, as suggested by ads juxtaposing the foster children and the territorial revival style home. Over dinner, Fia explains that her great-grandfather bought Crow Ridge Ranch in 1885, in the middle of the territorial period, and he "dreamed of building a home not just for his family, but for a community and for the brothers and sisters who believed in the teachings of the Pyre." The defining features of the territorial revival style, including the use of brick and pitched roofs covered in tin, distinguish it from the so-called

"Spanish," or "Pueblo" styles. While those other architectural styles are also associated with white settlement (the famous artist colonies and tourist centers of Santa Fe and Taos, for example), they represent settlement via the aesthetic appropriation of Indigenous/Mexican otherness, whereas the territorial revival style represents settlement by rejecting otherness and more transparently asserting white dominance. By turning a territorial revival style house into the setting of a white Christian death cult in the New Mexican desert, *The Unsettling* foregrounds the unsettling horrors of contemporary white Christian nationalism rooted in settler colonialism. When the teens attempt to escape, Fia threatens them with a gun, and, sounding like a paternalist slave master or a missionary chastising ungrateful slaves and servants, she exclaims "I brought you kids into my home, fed you, clothed you, put a roof over your head, and *this* is how you thank me?"

While numerous shows made during the Trump presidency represent MAGA villains, *The Unsettling* takes the unexpected turn of articulating Christian white nationalism to the foster care system. In addition to Fia, the show includes a character named Lorraine, Becca's case worker, who initially seems compassionate and well-meaning. When the foster children escape the compound and seek refuge in the cabin of a man who had previously escaped the cult, they call Lorraine for help. When she arrives, the foster care case worker stabs the man to death, reveals that she is also a cult member, and forces the children in her car so she can return them to the compound to be sacrificed. Loraine tells the children that the "system was never looking out for *any* of you. You were all just numbers, ghosts in the file cabinets that needed to be shuffled away." Their lives, she argues, can only be given a "purpose" by their deaths, making them "part of something bigger" (E7). *The Unsettling* thus promotes a revulsion to white Christian nationalism that oddly bleeds into its image of foster care. To be sure, popular depictions of foster care conventionally take the form of "horror stories," but they produce concrete local meanings in the context of TV production in New Mexico, where TV horrors of foster care seemingly rationalize Hollywood's siphoning of funds from social welfare.

At the same time, with its focus on DEI content and its patronage of industrial education and youth training, Hollywood in New Mexico projects an image of itself as an idealized private sector foster parent.

This is another, practical example of TV racial capitalism in place, where private enterprise takes the place of social welfare. State and industry boosters of New Mexican film production promote state subsidies with the language of corporate paternalism in which supposed jobs and other economic benefits implicitly take the place of, or obviate the need for, state spending on social welfare. As I have argued, however, such idealizations are belied by the small number of jobs created and the film industry's upward redistribution of wealth. Hence in terms of settlement and the appropriation of value from local Mexican and Indigenous people, we might say that film and TV studios in New Mexico constitute the contemporary equivalent of the territorial revival style of settler colonial nostalgia. Which is also to say that *The Unsettling*'s New Purity cult can be read against the grain as a displaced representation of the less sensational, slow violence of state subsidies paid to Los Angeles-based production companies and finance capitalists.

Atlanta, mostly filmed in the title city, mixes multiple genres, but several episodes incorporate conventions from horror. For example, "Terry Perkins" (S2E6) is a Black revisioning of *Whatever Happened to Baby Jane?* about homicidal elderly brothers in an old Atlanta mansion; in "Woods" (S2E8) Al "Paper Boi" Miles is chased through the forest by an enigmatic man with a knife; "Crank Dat Killer" (S4E6) depicts a serial killer targeting people who have recorded internet videos dancing to the Soulja Boy song "Crank That"; and finally, the topic of this section, an episode titled "Three Slaps" (S3E1) features foster parent serial killers.

Based on a real incident, "Three Slaps" tells the story of Loquareeous Reid, a Black middle school student wrongly taken from his mother by Child and Family Services and then adopted by a white lesbian couple, Amber and Gayle, who are already the foster parents of three other Black children.[7] Unable to spell his name, Amber changes it to "Larry." The foster parents serve the children raw chicken, so they are always hungry. They are also forced to work in a garden and to sell kombucha at farmers' markets. When a social worker visits to check on the children, Gayle kills her. With the pretext of visiting the Grand Canyon, the two women hit the road with the children, arriving one night at a bridge-covered lake illuminated by streetlamps. In an attempted murder/suicide, just before the bridge the foster parents swerve off the road and into the lake, but not before Loquareeous jumps out of the vehicle to safety (a subsequent scene in the episode reveals that the other children were rescued). Loquareeous witnesses the van drive into the water, and we hear the splash of the impact. What follows is a haunting shot/reverse shot sequence establishing a frightening relationship between the water and the young Black witness. A shot looking up at the illuminated bridge, from the perspective of the water, as Loquareeous approaches the bridge's railing, is followed by a shot from behind Loquareeous as he looks down from the bridge into the lake. The ghostly light of a single streetlamp, combined with the sound of crickets on the soundtrack, evoke horror, as if the character is in danger of being pulled into the water.

The scene of Loquareeous looking down into the lake is foreshadowed by the episode's cold opening, represented as a nightmare from which Loquareeous awakes. The episode begins with a nighttime aerial shot of the same bridge with the light of a small boat visible in the water where two men, referred to in the credits as "Black" and "White," drink beer and fish. When Black says, "this place always gave me the heebie-jeebies.... I almost drowned in it once when I was like eight.... I don't know what it was man, but I felt like I was being pulled." White responds that he probably *was* being pulled into the water, since the lake was on top of a Black town that was buried when "the state government built a damn and flooded the place." Black: "So, there are Black people under us right now?" White: "Yeah, why do you think so many people die around her every year.... It's haunted.... Lot of souls down there. That's what pulled you under." The scene ends when six ghostly Black hands pull Black from the boat and into the water.

Atlanta here condenses the history of Black displacement and appropriation in Georgia. Reviews of the episode note that the opening is inspired by the history of Lake Lanier, produced in 1956 when the US Army Corps of Engineers damned the Chattahoochee River. The water covered a valley of trees and mostly white-owned farmland. The whiteness of the valley under Lake Lanier is the result of an earlier moment of violent racial cleansing in 1912 when, in the wake of the lynching of a Black man and the public execution of two Black teens who were accused of raping a white woman, white nightriders attacked the Black settlement of Oscarville and drove out over a thousand Black residents, burned homes and churches, and appropriated crops, livestock, and ultimately land. Forty-four years later, what remained of Oscarville was submerged under Lake Lanier.

By starting at and returning to Lake Lanier, the episode suggests that the white foster parents who attempt to murder their Black foster children are part of the longer history of state-sponsored anti-Black violence. "Three Slaps" resembles Derrick Bell's critical race theory speculative fiction "The Space Traders," which suggests that historically, the US government has sacrificed Black interests whenever they conflict with white interests. The *Atlanta* episode also indirectly suggests a subtle critique of location shooting in Georgia. State subsidies that divert resources from poor people of color to Hollywood build on the longer histories of violence and displacement represented by a location like Lake Lanier. *Ozark*'s lake scenes, for example, were set in Missouri but filmed in Georgia at Lake Allatoona and Lake Lanier. The lakefront home where the white, money-laundering Byrd family live is in fact on Lake Lanier, which was named after Sydney Lanier, a Confederate veteran of the Civil War. As I noted in chapter 2, several studios in Georgia are located near waterways and include housing developments surrounding artificial lakes. Bouckaert Farms, for example, the massive 8,000-acre development where scenes from *The Avengers: Endgame* and *Black Panther* were shot, advertises its miles of Chattahoochee River frontage and its lake. Tony Stark's lakeside cabin in *The Avengers: Endgame* was filmed there. (It is available to rent on Airbnb.) A planned housing development at Bouckaert Farms, designed by a former vice president for Walt Disney Imagineering, further encourages settler colonial fantasies by selling homes and plots of land called "farmettes."[8]

In contrast with such corporate boosterism, *Atlanta* represents Lake Lanier as a space of Black gothic horror. At the same time, the homicidal paternalism of the white foster parents suggests a sensational representation of Hollywood paternalism and its overblown promises of good jobs. The potentially critical representation is compromised, however, by the fact that the producers of *Atlanta* also benefit from state film subsidies that redistribute wealth upwards. Viewed from that perspective, critical representations of state welfare systems promote the material interests of Black creatives who profit by appropriating public resources.

Mostly set in Houston, Texas but filmed in Atlanta, the limited series *Swarm*, created by *Atlanta*'s Donald Glover and Janine Nabers, depicts a queer Black former foster child who becomes a serial killer. Along with Glover, Nabers is a key figure in Georgia TV production, serving as a writer and producer on *Atlanta* and *The Watchmen*. *Swarm*'s protagonist, Andrea Green, or Dre, is an obsessive fan of a Beyonce-like performer named Ni'jah. That obsession bonds them to their upper class foster sister, Marissa Jackson, who commits suicide in the first episode when her boyfriend cheats on her. Dre proceeds to murder not only that boyfriend, but also the abusive boyfriend of a co-worker, and other Black people who insult Ni'jah on social media. While in proceeding episodes Dre dresses and styles themselves in often eccentric but recognizably "feminine" ways, in the final, Atlanta-set episode they go by the name Tony and adopt a "masculine" self-presentation. As Tony, Dre murders their girlfriend Rashida, an upper class Black graduate student, also for insulting the singer.

Glover explains that when directing Dominique Fishback, who plays Dre, he and Nabers "specifically didn't even let her know her own backstory. I felt like if she knew it, she would be telegraphing it unconsciously to the audience. I didn't want us to understand why Dre was doing things."[9] A possible explanation for that decision is revealed in episode 6, where a police detective investigating the murder visits Dre's case worker at the Department of Family and Protective Services in Houston. The case worker refuses to provide any information about their background, angrily saying that "the only reason you want Andera's sob story is so you can absolve yourselves. You need there to be a reason she was so messed up, so you don't have to sweep up your own front door and realize you are just as flawed."

The withholding of Dre's backstory, however, can also be interpreted as another kind of "sweep up" that seemingly rationalizes *Swarm*'s appropriation of public resources from welfare systems that supposedly make monsters. As Glover told a reporter, he advised Fishback to "Think of it (Dre) more like an animal and less like a person." He continued that the character "reminds me of how I have a fear with dogs because I'm like, *you're not looking at me in the eye; I don't know what you're capable of.* That's what I wanted to have with her," and Fishback "really nailed that."[10] Dre's animality is reinforced visually in the shocking conclusion to the pilot. After they bludgeon their first victim to death, the show's producers insert a shot/reverse shot sequence in which Dre seemingly exchanges a glance of recognition with the murdered man's Rottweiler. In other words, *Swarm* animalizes its protagonist, a character from the kind of race/gender/sexual/class background that would most benefit from public monies diverted to subsidize TV production in Georgia.

This naturalization of working-class Black evil, produced by more privileged Black creatives, legitimates class hierarchies. Contributing to this, *Swarm*'s set design suggests a condescending perspective on a working-class character. In episode 1, Dre leaves the poor apartment they share with their foster sister and visits a market with bars on the windows and graffiti on its sign, where a cashier sells beer, wine, and junk food from behind bulletproof glass. Episode 2 sees Dre living in a cheap motel with laundry drying on the second-story railing. Their small room features stained walls, a dirty mirror, and a thick blanket of fast food remains, including chicken wing bones crawling with ants. Such settings screen as class porn that aestheticizes poverty.

Swarm represents the living spaces of Dre's victims in dramatically different ways. While they also murder a few working-class people, including a stripper and a tow truck driver, Dre has a proclivity for attacking wealthy people, especially wealthy Black people. In episode 3, for instance, they kill a Black man living in an "open concept" condo featuring a modern kitchen with a granite island, a minimalist white living space decorated with colorful art and furniture, a swimming pool, and a large walk-in closet, where Dre attacks him with a hammer. (See Figure 4.1.)

When Dre unsuccessfully attacks their Black former foster parents, we learn that they live in a large, Italianate home with a grand foyer and a three-car garage. Finally, while Dre sleeps in a van with a knife on their pillow, Rashida lives in a "tastefully" decorated, plant-filled apartment at the top of a grand staircase.

As previously suggested, junk food is everywhere in *Swarm*, serving as props that anchor Dre in an upper-class vision of poor Black people. As Detective Lorretta Green declares in episode 6, "it all started with junk food." Crime scene photos picture a crumpled Flaming Hot Cheetos bag and Krispy Creme, Taco Bell, and Popeyes packages. After murdering the condo-dweller in episode 3 and finding to their disappointment a refrigerator filled with fresh fruit and vegetables, Dre instead takes the bags of chips and boxes of Cheez-Its from his large pantry. Similarly,

Figure 4.1: Serial killer Dre cleans up a murder scene in the home of her upper-class victim on *Swarm*.

they break into the apartment of a sound tech for Ni'jah and find only "healthy" food there. Dre ultimately seduces the man with powdered donuts and barbeque potato chips, the remains of which are depicted strewn across his bed in the morning. Junk food even seeps into the show's visual palette. *Swarm*'s cinematography emphasizes saturated colors of red, blue, purple, and yellow that mimic junk food packaging. The show employs an entire junk-food aesthetic as a kind of filter for seeing poor Black people with poor diets. *Swarm* in this regard bares comparison with *Being Mary Jane*, which adopts a similarly condescending upper-class Black perspective on working-class Black characters (chapter 6).

Cultural critic Bertrand Cooper argues that contemporary middle- and upper-class creatives in film and TV materially benefit by representing poor Black people for white audiences. Such creatives, they argue, make careers out of representing poverty they haven't experienced, thereby obscuring class differences, and projecting a collective "we" which allows them to represent and speak for poor Black people. Cooper argues that the 2020 police murder of George Floyd led to the greenlighting of a host of Black films and TV shows. But in practice, more privileged Black creators gained opportunities denied to Floyd, who "would not be hired to work within any of the institutions which now produce popular culture in his honor because he never obtained a bachelor's degree." Cooper cites Glover, from a middle-class background, as an example. Discussing the success of *Atlanta*, Glover said "as a Black person, you have to sell the Black culture to succeed." The show, Cooper points out, "draws heavily on poverty, policing, prison, violence, rap and the culture of the Black poor. It's not the environment Glover grew up in or the culture he practiced in high school or during his 20s or 30s, but it *is* the one he sold to gain two Emmys."[11] *Swarm*'s all-Black production team includes other similarly privileged Black creatives, including Julliard educated co-creator and showrunner Nabers, Glover's brother Stephen, and Malia Obama, who helped write episode 5, where Dre breaks into their foster parents' home. (Barack Obama is reportedly a fan of *Atlanta*.) But in contrast with the scenarios analyzed by Cooper, in which more privileged creatives obscure class differences among Black people by appropriating experiences of Black poverty, *Swarm* bluntly reproduces class hierarchies with its depiction of a former foster child as an othered animal or an "it," in Glover's words. Ironically, in media interviews, Nabers celebrates the show as a contribution to diversity not only in terms of Black TV production talent, but also in terms of content: "One of the things for me as a storyteller is just allowing Black women to not be placed in a box of, for lack of a better word, 'Blackness.' We're not a monolith. Black people are so many different things. With *Swarm*, we really wanted to create a character that just felt so unique."[12] Nabers here redescribes a show based in class hierarchies among Black people in DEI terms, such that the representation of a poor Black serial killer supposedly undermines assumptions of a monolithic Blackness. Although the concept of racial capitalism is often imagined in terms of white elites who profit from anti-Blackness, building on Cedric Robinson's analysis of more privileged early twentieth century Black filmmakers targeting working class Black audiences (chapter 1), I argue that shows such as *Atlanta* and *Swarm*, as well as many Black melodramas made in Atlanta (chapter 6), are products of Black racial capitalism in place.

Supernatural DEI Horror

Supernatural horror shows also incorporate DEI elements while indirectly referencing their participation in the violence of upward wealth

Figure 4.2: Having killed one white supremacist, the vampire Lemuel (a former slave) prepares to dispatch a second on *Midnight, Texas*.

redistribution. Representations of supernatural horrors mediate the violence of TV racial capitalism in place. The NBC series *Midnight, Texas* (2017–2018) is perhaps the most "DEI" of all the horror shows in this chapter. It is based on a series of novels by Charlene Harris, who also wrote the popular books that inspired the HBO series *True Blood*, which figured vampires as "minorities." *Midnight, Texas* adopts a similar perspective on the diverse, supernatural denizens of the titular town. Lemuel is a Black Vampire, whose white human lover and hit woman is Olivia. (See Figure 4.2.) Latinx minister Emilio Sheehan transforms into a "weretiger." Joe Strong is a white fallen angel and Chuy is his half-demon Latinx husband, while Fiji Cavanaugh is a Black witch and Bobo, her white boyfriend, is disowned by his racist family. The small Texas town is a precarious haven for supernatural and human outcasts, under siege from both the police and a white-supremacist biker gang called the Sons of Lucifer. By costuming them in vests with side-by-side patches of the Confederate flag and the flag of Texas, the show reminds viewers of the state's founding in and history of anti-Black and anti-Mexican racism (S1E2). Those associations are reenforced by the set design of the Sons' unofficial club house, a country and western bar where cowboy hat and boot wearing patrons line dance while members of the gang play pool in a backroom decorated with a large Confederate flag and a neon sign in the shape of the state of Texas (S1E4).

The construction of Midnight, Texas as a haven for diversity, equity, and inclusion is at odds with the places where the show was shot. Interiors were filmed on soundstages at Netflix's Albuquerque Studios, and the show reused a backlot church to represent the Latinx weretiger's church. As detailed in previous chapters, Albuquerque Studios is part of an exclusive, state-subsidized, mixed-use residential development, practically and ideologically invested in carceral institutions that criminalize poor Mexicans and Indigenous people in Albuquerque's South Valley (chapters 3 and 7). Many other scenes in *Midnight, Texas* were filmed in nearby Las Vegas, New Mexico, where the modern western TV show *Longmire* was also shot. But whereas *Longmire* employed the old Spanish Colonial-style Las Vegas Plaza, *Midnight, Texas* chose a different local location, the Railroad Avenue Historic District, named for an 1899 Spanish Mission Revival train station and Harvey hotel. The hotel is used to represent the restaurant where humans and supernatural beings work, drink and eat, while the turn-of-the-century buildings across the street are employed for exterior shots of where the show's central characters live. In its own kind of revival, the show's use of the location suggests a nostalgia for industrial railroads as the precursor to the contemporary film and TV industry in the region. (See chapter 2's analysis of the Albuquerque Rail Yards, where *Midnight, Texas* also filmed.) Read critically, the implicit comparison suggests the continuities between the two industries as engines of inequality. The expansion of the Atchison, Topeka, and Santa Fe Railroad in the area encouraged white squatters and land speculators who appropriated Indigenous and Mexican lands. In 1890, Mexicans took organized direct action in the form of a group called Las Gorras Blancas, which destroyed fences and burned barns and railroad ties. These events haunt *Midnight, Texas* in the sense that its depiction of white violence leaves unrepresented and unacknowledged the region's actual history. At the end of the nineteenth century, railroad boosters promoted "progress," as do today's TV producers in New Mexico. But neither industry has benefited poor and working class Mexican and Indigenous people there. On the contrary, both enterprises have depended on the appropriation of land and resources. Today, the Las Vegas population is almost 80% Latinx, with a per capita income of

$23,000 a year, and over 30% living below poverty (whereas the national average is 11.6%). The omnipresent threat of violence on *Midnight, Texas*, I argue, indirectly represents the slow violence of film productions that redistribute wealth upwards and treat poor communities as disposable.

Season 1 of *Preacher*, filmed in and around Albuquerque, likewise treats rural New Mexican communities as disposable. It features a brooding minister named Jesse Custer (Dominic Cooper) as the titular preacher in the fictional town of Annville, Texas, where he battles evil both human and supernatural, alongside his lover Tulip (Ruth Negga) and his vampire friend Cassidy (Joe Gilgun). *Preacher* diversifies its source material by casting a Black actor as Tulip, who is an Anglo character in the graphic novel on which the show is based.[13] The program presents critical perspectives on factory farming, represented by the Quincannon Meat and Power Company, owned by the odious villain of season 1, Odin Quincannon. In one episode, he displaces a Mexican couple, bulldozes their farmhouse, and turns their land into pasture. Quincannon is also pictured building elaborate models of the Alamo in his office while enjoying the sounds of the slaughterhouse on an intercom and holding a baby made from ground beef stuffed into a toddler's jumper. The show's evocations of frontier violence, land dispossession, and US empire through the character of Quincannon represents an implicit indictment of racial capitalism on the register of narrative content, with the character's sadistic rapacity figured as even more monstrous than *Preacher*'s vampires and demons.

But the progressive potential of *Preacher* is undermined by the pleasure the program encourages viewers to take in scenes of spectacular violence that revel in the grittiness of the show's impoverished locations. Many such scenes are set in the Sundowner Motel, which also gives its name to episode 5 of season 1. (See Figure 4.3.)

In one scene, for example, two angels track Jesse to the seedy motel, where they are followed by a third angel who attacks them. The angels are immortal so as they shoot, stab, and club each other to death, they are reincarnated in new bodies and the battle continues as the corpses pile up. Those scenes were filmed at the actual Sundowner Motel on Albuquerque's Central Avenue. As the name indicates, Central Avenue was once a prosperous street, a stretch of Route 66 that ran the length of the city and featured numerous restaurants, motels, and tourist attractions.

Figure 4.3: A Latina maid pushes a cleaning cart at Albuquerque's Sundowner Motel in *Preacher*.

But it fell on hard times when Route 66 was superseded by Highway 40. While parts of it remain relatively prosperous, Central Avenue's southeast portion, where *Preacher* was shot, is filled with boarded-up businesses and desperately poor Indigenous, Mexican, and Black residents.

Preparing to visit the area while researching this chapter, I packed a camera, but when I got there, it felt wrong to take pictures; the prospect struck me as a kind of extractive voyeurism that echoed TV's racial capitalist aesthetic. But that doesn't seem to have troubled the producers of *Preacher*, who used New Mexico tax dollars to shoot slumming scenes of body horror at the Sundowner, where the upward redistribution of wealth and racialized labor exploitation that help make such programs possible are an implicit—but invisible—background to the show's supernatural violence. Audiences are called upon to enjoy such representations of spectacular violence while shielded from knowledge of the more everyday violence of TV racial capitalism in place and the theft and exploitation it organizes.

The scenes of apocalyptic violence that conclude *Preacher*'s first season spark a similar interpretation. The methane produced by Quincannon's cattle explodes in the final episode, destroying Annville and killing all its inhabitants except for Jesse, Tulip, and Cassidy. For the next two seasons, the series moved both its setting and shooting location to Louisiana, and the three protagonists never mourn and hardly think about the town and the death of their friends and neighbors. The season 1 finale ends with images of genocide partly produced in an impoverished location. The establishing shots depicting Annville were filmed in the small New Mexican village of Estancia. Most of the local businesses depicted are closed, and their incorporation in the program aestheticizes rural poverty. The images of rural poverty in *Preacher* resemble those in *Midnight, Texas*, where the small section of Las Vegas, New Mexico used to represent the town is filled with crumbling old buildings.

Shows shot in poor places often depend on the existence of "picturesque" poverty in locations where Indigenous people and people of color live. Estancia is over 50% Latinx, with more than 25% of its population living below the poverty line (over twice the national average). The largest employer is a private prison that detains migrants from Mexico and Central America on behalf of ICE. While Estancia's tax dollars subsidize media production in the state, there are no signs of trickle-down benefits for its residents. Although *Preacher* aligns itself with diversity, Annville's fate brings into sharp relief the local impact of New Mexico's tax incentive program, which provides an extra 5% "Uplift Zone" incentive for films and TV shows shot in rural areas. The show renders Annville and its inhabitants disposable, while its protagonists move on and live on. The narrative thus anticipates how *Preacher* lived on, moving to New Orleans while abandoning Estancia, disposing of the village and its people.

Sleepy Hollow (2013–2017) similarly combines DEI casting and narratives, on the one hand, and representations of disposability on the other. It draws from Washington Irving's "The Legend of Sleepy Hollow" and "Rip Van Winkle" to represent Ichabod Crane's transportation from the eighteenth century to present-day New York and Washington, DC, where he works to prevent the Apocalypse with Abbie Mills, a Black sheriff who, over the course of the series, becomes an FBI agent. The pair is aided by Abbie's sister, Jennifer, who hunts supernatural relics. The first two seasons were shot in North Carolina but the next two were filmed in the suburbs of Atlanta, with Georgia trees standing in for the forests of Sleepy Hollow. While among scholars and activists the police are understood as agents of structural racism, *Sleepy Hollow*, like so many other shows, sees them though a heroic DEI lens in which women of color police fight evil (in this case of the supernatural kind). Such depictions are influential, representational components of a police and prison televisual complex that shores up the police as they support and protect TV's upward redistribution of wealth (chapter 3).

Similarly, in season 4 it is revealed that the US President is an unnamed Black woman. The existence of a Black woman president in the present projects an image of progress, even though the episode was produced in the year after the election of white nationalist President Donald Trump. In the season finale, an evil corporate head named Malcolm Dreyfus orders the four horsemen of the apocalypse to take the President hostage but Crane and Thomas save her. Much of season 3 focuses on Crane's efforts to become a naturalized US citizen, and in the season 4 finale, the first Black woman president makes him one, saying "welcome to America." The scene is shot in a sentimental style, with warm, low lighting and romantic string and piano music, suggesting that the program's narrative arch culminates in a patriotic spectacle.

As that conclusion suggests, *Sleepy Hollow* deploys diverse casting and DEI narratives in the service of US nationalism.[14] It is derivative of *National Treasure*, an action/adventure film based in fantasies about the hermeneutic codes the founding fathers supposedly incorporated into documents and other artifacts. The film inspired a recent DEI update, a TV show filmed in Santa Fe, New Mexico, titled *National Treasure: The Edge of History*, about a young Mexican woman Dreamer who dreams of becoming a citizen so she can work for the FBI. To return to *Sleepy Hollow*, the show's multiple flashbacks infuse revolutionary war figures with all the excitement and thrills of supernatural hermeneutics. Icons of US patriotism such as George Washington, Ben Franklin, Paul Revere, and Betsy Ross are transformed into cryptographers, alchemists, and wizards. The program thus promotes US nationalism even in the face of ongoing inequalities, a reality *Sleepy Hollow* participates in via the diversion of public funds to cover the show's production costs.

Sleepy Hollow's nationalist DEI vision is further undermined by its labor practices. Nicole Beharie claims that she and Tom Mison, the white, English actor who played Crane, became sick during the filming of season 3 but they were treated differently by the show's producers. He was able to take a month to recuperate in England whereas she was pushed to keep working, until she faced a medical crisis. When she complained, her character was written out of the show, and Beharie claims she was subsequently blacklisted as "problematic." Drawing on interviews with cast and crew, journalist Maureen Ryan has demonstrated that despite its supposed inclusiveness, the writers' room and set were beset with casual racism.[15] At the same time, the actor who played Abbie's father also lost his job. This was after the Black actor who played a police chief in season 1, Orlando Jones, was fired. As Beharie concludes, "everyone of color on that show was seen as expendable and eventually let go."[16] The expendability of Black actors is reproduced in the show's narrative by the disposability and interchangeability of actors of color. In the character's final appearance on the show, Abbie saves the world from demonic destruction by sacrificing herself to supernatural forces. The final season of *Sleepy Hollow* replaced Beharie's character with a new woman of color partner, Homeland Security Special Agent Diana Thomas (played by South Asian actor Janina Gavankar), and a new group of multicultural allies for Crane. The centering of white actors and the disposability/interchangeability of actors of color is consistent with the broader political economy of TV production in Georgia, which helps make poor people disposable by appropriating public resources. As this example suggests, TV racial capitalism in place uses DEI representations to legitimate raced and gendered labor exploitation, both in casting and in the larger low-wage labor economy on which state subsidized TV production depends.

Although not idealistically nationalist like *Sleepy Hollow*, *Lovecraft Country* uses horror to align the show with a heroic history of the African American civil rights movement. Largely set in 1950s Chicago and the fictional town of Ardham, Massachusetts, most of *Lovecraft Country* was filmed outside Atlanta. The show draws on the work of the infamously racist horror writer H. P. Lovecraft to reflect on the history of anti-Black racism, particularly in the northern United States. The story of a Black family in an epic battle against wealthy white supremacist magicians includes numerous representations of segregation, racist police, and vigilante violence. *Lovecraft Country*'s premiere episode starts with Tic on a segregated bus, while in the final third he joins others on a road trip where they confront segregation and racist violence. Tic's uncle, George Freeman, is the author of a *Green Book*-like guide for driving while Black, recording where Black people are welcome and unwelcome.

Figure 4.4: Tic and Leti in the forest as a racist policeman approaches in *Lovecraft Country*.

Killing two birds with one stone, Leti and Tic travel to Massachusetts, and George drives them while updating his guide. Along the way they encounter a segregated movie theater and ice cream parlor in scenes that recreate famous anti-segregation photos by Gordon Parks, imbuing *Lovecraft Country* with the gravity of that history. They stop in one place so George can check on a restaurant owned by a Black woman, only to discover that she had been burnt out by local racists, who proceed to fire on the three characters and run them out of town. The episode's title, "Sundown," refers to the history of so-called sundown towns which required Black people, under threat of violence, to leave before sunset. Leti, Tic, and George are detained on a forest road by a racist sheriff who tells them they have only minutes to get out of the county. (See Figure 4.4.)

In a tense sequence, the cop tails them as they head toward the county line. They make it there in the nick of time, only to be roadblocked by the sheriff and his deputies, who are seemingly on the verge of lynching the three characters in the forest when they are attacked by "shoggoths," vicious creatures with claws, hundreds of teeth, and eyes all over their gelatinous skin. As the showrunner and writer Misha Green notes, she used Lovecraft's monster as "a metaphor for the racism that's always been running through America and globally," while Majors adds that "white racists, or racists in general, are that much more terrifying than a shoggoth."[17]

Lovecraft County, in other words, critically represents the historic limits to Black mobility, including with fantastic counter-representations of the character Hippolyta Freeman. Married to George, she comes to lament that while he gets to travel, she stays put, so Hippolyta embarks on her own road trip. Her longing for freedom of movement is represented by her own driving, and her encounter on the road with a motorcycle-riding Black woman. She ultimately arrives at a magic portal which transports her into outer space; to 1920s Paris, where she is one of Josephine Baker's dancers; and finally, to Africa, where she leads a group of Dahomey warrior women in battle against Confederate soldiers.[18] In terms of narrative and imagery, *Lovecraft Country* makes important interventions into the politics of representation. As Green argues, the show uncovers histories that are "buried . . . when we center a narrative just around whiteness."[19] In addition to shedding light on marginalized histories and Black freedom dreams, *Lovecraft Country*, according to Green, reclaims "genre space for people of color, for people who have been left out of this space."[20] Green here describes the show's achievements in terms of inclusion—more inclusive historical representations, more inclusive visions of genre.

Inclusion, however, looks different when we shift from representation to consider the political economy of TV production based in the upward distribution of resources from poor people of color to more privileged Black creatives. From that perspective, we could read Hippolyta's mobility as an allegorical representation of class mobility for Black TV producers. But what of the immobile? How might *Lovecraft Country* impact less privileged people who live where it was filmed? Some scenes were shot on location, but many more were filmed on soundstages at Blackhall/Shadowbox Studios, including Hippolyta's portal travels. Many shows in the Marvel universe are also filmed on Georgia soundstages, enabling the creation of fantastic worlds, physically and imaginatively segregated from the communities surrounding them. As noted in chapters 2 and 3, Blackhall/Shadowbox Studios is located near low-income Black neighborhoods and has become the target of protests over its privatization of a public forest park along the South River. The studio where *Lovecraft Country* was filmed, activists argue, has appropriated one of the few natural spaces available to Black people in the area. Blackhall/Shadowbox is also adjacent to the planned Cop City, which will train police to work in Black communities. Both institutions are on public lands, alternatively converting the land's value into TV shows and policing, in a broader context where TV and the police work in tandem, the one legitimating the other, and vice versa.

Finally, Blackhall/Shadowbox deprives people of access to natural beauty and its regenerative powers by privatizing what was until recently public parkland forests and waterways. By contrast, *Lovecraft Country* employs forest land near the studio and in Illinois to produce images of racist terror and supernatural violence. Scenes of lynch mob police, shoggoth attacks, and the frenetic efforts of Leti, Tic, and George to escape both, were all filmed in forests. When the trio safely arrives in Ardham, they discover that the evil, white supremacist Sons of Adam Lodge is surrounded by dense forests. To be sure, forests are conventional horror settings, but they have special resonance in Black contexts as trees have been instruments of lynching. As Joseph Lewis argues, in *Lovecraft Country*, "monstrous encounters in natural settings like the forest offer a historically and culturally specific narrative that the threat of violence is ever-present, as the Black character traverses the terrain of the countryside."[21] The prospect of Cop City surrounded by forests is only the most recent reminder of such contexts. But in the history of TV production in Georgia, forests are perhaps overrepresented, largely because of *The Walking Dead*, where many gruesome struggles between zombies and humans take place, but also because of *Sleepy Hollow*, with forest scenes filmed in the same area as those in *Lovecraft Country*. In the case of *Lovecraft Country*, and the episode of *Atlanta* titled "Woods," in which one Black character chases another through the forest with a knife, the use of forests as settings for terror specifically underlines the claim by local people that the TV studio had appropriated a public space of Black recreation. Although *Lovecraft Country* presents critical images of the historic segregation of public places, by filming at Blackhall/Shadowbox it supported the privatization of nature and the resegregation of public parks.

Return of the Repressed, Repression, and DEI

Robin Wood, in his influential essay "An Introduction to the American Horror Film" (1978), famously distinguished between progressive horror films from the early 1970s representing the return of the repressed in the form of attacks on heteronormative families, on the one hand (*Sisters*, *Night of the Living Dead*, *Texas Chain Saw Massacre*), and reactionary films that promoted gender/sex repression, on the other (*The Omen*, *Halloween*, and *Alien*).[22] One doesn't have to agree with the categorization of particular films to find Wood's framework useful. Taken together, for example, the horror shows in this chapter combine the return of the repressed and the promotion of repression, often in the same TV program.

Return of the repressed. The accumulation of angel corpses in *Preacher* represents the return of a repressed awareness of the violence of film and TV subsidies. The ghostly Black hands that pull "Black" into the lake in the "Three Slaps" episode of *Atlanta* figure the return of repressed local histories of anti-Black violence, including the violence of wealth redistribution in the local TV production. *Swarm* suggests the return of poor Black people repressed by boosters of film incentives and the diversion of public money from social welfare to Hollywood. The denizens of *Midnight, Texas* represent the return of those repressed by white supremacy, while *Sleepy Hollow*'s supernatural eruptions suggest the return of the repressed contradictions and exclusions in US national history.

Promotion of repression. Depictions of a monstrous foster care system in *The Unsettling*, "Three Slaps," and *Swarm* encourage the repression of social spending; with its genocidal vision of the destruction of Annville, *Preacher* performs the repression of Latinx people under conditions of TV racial capitalism in place. By reifying evil in the form of a working-class queer Black serial killer, *Swarm* promotes police repression and incarceration. And finally, *Sleepy Hollow* promotes heroic police and the disposability of Black actors. The promotion of police repression is one representational wing of the police and prison televisual complex.

But there is a third horror effect not anticipated by Woods—forms of DEI that substitute for material inequalities in TV production and that justify prisons and the police. While devoted to provoking pleasurable fears and anxieties among viewers, horror shows also employ DEI representations as a way to contain, dissipate, or dissolve troubling knowledge about how TV production diverts life-sustaining resources from poor Black, Latinx, and Indigenous peoples while promoting the police and prisons that target them.

5

Science Fiction TV

The DC superhero show titled *Doom Patrol* (HBO Max/Max, 2019–2023) begins by reducing Lawrenceville, Georgia to rubble. Lawrenceville is the location of the infamous Gwinnett County Prison used to film crime shows (chapter 3). *Doom Patrol* uses Lawrenceville to play the fictional town of Cloverton, Ohio, home to Doom Mansion, owned by a scientist in a wheelchair named Niles Caulder, and sheltering a group of misfit superheroes. In the pilot episode, the historic Lawrenceville Square is used to shoot an attack on Cloverton by Caulder's enemy, the supervillain Mr. Nobody. As the Doom Patrol helplessly look on, Mr. Nobody uses his powers to open a large portal next to the plaza which swallows up the entire town and its inhabitants, including Caulder.

This scene recalls *Preacher*'s destruction of a rural New Mexico community (chapter 4), but with a notable twist given *Doom Patrol*'s location: the Lawrenceville Square is decorated with a Confederate monument, which is visible in the show. Whereas most Confederate monuments were erected decades earlier, Lawrenceville's memorial dates to the Bill Clinton era. In 1993, the Daughters of the American Confederacy successfully petitioned the city council to allow them to install a large granite memorial with a bas relief sculpture of a Confederate soldier in front of the square's courthouse. It was finally placed in storage in 2022 but was the subject of local controversy when the *Doom Patrol* scenes were filmed. In that context, the show visually obliterated the square with its monument to white supremacy. Although the toppling of a Confederate memorial might not have been intentional for *Doom Patrol*'s producers, this example raises the question of how MAGA-era TV shows responded or not to racism, white nationalism, and settler colonialism in the poor places where they were made.

Georgia and New Mexico have hosted extensive science fiction productions, including many DC and Marvel films and TV programs. Greater Atlanta hosts *Watchmen* (HBO, 2019), *Raising Dion* (Netflix,

2019–2022), *Naomi* (The CW, 2022), and *Black Lightning* (The CW, 2018–2021), in which Black superheroes fight evil, especially anti-Black racism. Partly because of the history of Area 51 and stories of alien sightings, New Mexico has attracted TV shows about alien phenomena, as in *Roswell, New Mexico* (The CW, 2019–2022) and *Outer Range* (Amazon, 2022–), science fiction westerns shot in Santa Fe and Las Vegas, New Mexico. Both kinds of science fiction programs employ diverse casts and behind-the-scenes talent, and they critically engage topical political issues like police violence, white nationalism, Confederate memorials, settler colonialism, and migrant detention. Set in speculative worlds, science fiction shows are nonetheless inspired by their contemporary social and political contexts, including their conditions of production. While to different degrees the programs in this chapter all present challenging critical content about white nationalism, they represent the political economy of their production differently, with most displacing and obscuring it while a few directly represent it.

The Black superhero shows were filmed amidst the sea of Confederate monuments in and around Atlanta, and their production was subsidized by white nationalist state officials seeking to protect them from being toppled. The state guardians of Confederate memorials such as Governor Brian Kemp (2019–) are also film subsidy boosters, suggesting a connection between white nationalism and state budget priorities that divert funds from health care, education, and low-income housing for poor Black and Latinx people to Hollywood TV producers. *Watchmen*, *Raising Dion*, and *Naomi* respond to the Confederate memorial/TV subsidy nexus on which they depend by excluding from the screen the monuments to white supremacy in their Georgia locations. Although not directly represented, however, Confederate memorials haunt the shows, reappearing in transmuted forms—as KKK supervillains (*Watchmen*), a giant white monster (*Raising Dion*), and even Superman (*Naomi*)— that imaginatively disconnect the show's content from the Confederate memorial/TV subsidy nexus. By contrast, *Black Lightning* represents a form of Black feminist iconoclasm, in which a Black woman superhero explodes a Confederate memorial, drawing attention to what the other series hide.

The two New Mexico-made alien programs partly differ in their responses to their conditions of production. Set in southeastern New

Mexico but filmed in Santa Fe and Las Vegas, *Roswell, New Mexico* employs a diverse cast and writers' room to make episodes about anti-immigrant white nationalism and migrant detention, topics that were part of national discourse but with special resonance in Las Vegas and Santa Fe, where the show was shot, in local landscapes blighted by film-studio adjacent prisons and migrant detention facilities. At the same time, and in conflict with its representations of anti-migrant policing, *Roswell, New Mexico* aestheticizes and romanticizes police and prisons, including the real New Mexico prisons used for its settings and backdrops. The show in this way disavows its complicity in the police and prison televisual complex, or the set of mutually beneficial relationships between TV production, prisons, and the police, including the use of police departments and prisons as filming locations; the use of police equipment as props; the employment of prison staff and police as security guards, actors, and advisors; and public relations spectacles in which the prisons/police legitimate Hollywood and vice versa. (See chapter 3.)

Like *Roswell, New Mexico*, *Outer Range* also seems based in disavowal, but in the form of modernist complexities and ambiguities that resist interpretation. The performance by Indigenous actor Tamara Podemski (Anishinaabe descent) as Sheriff Joy Hawk, however, pushes beyond disavowal to ground the show in its settler-colonial preconditions of production. Podemski's interpretation of the role suggests that the DEI demands of TV casting and characterization abstract shows from the world around them, including where they are made. Ultimately, her performance projects the here and now of Indigenous survivance in the face of ongoing settler colonialism, including the televisual kind.

Georgia's Black Superheroes

As in other TV genres, Black superhero shows present critical reflections on MAGA white nationalism. After a white supremacist shot nine Black people at a Bible study meeting in Charleston, North Carolina's Emanuel African Methodist Episcopal Church in 2015, critical attention and action focused on Confederate monuments. Filmmaker and activist Bree Newsome responded to the shooting by climbing a flagpole in front of the Charlotte, North Carolina state house and removing a Confederate flag. Newsome's protest was widely covered in various media, and

Figure 5.1: Sister Night from *Watchmen*.

she appeared on a several TV talk shows. In 2017, she wrote a *Washington Post* opinion piece titled "Go ahead, topple the monuments to the Confederacy. All of them," condemning the violent white supremacist "Unite the Right" rally protesting the proposed removal of a statue of General Robert E. Lee.[1] Images of Newsome, including art representing her as a superhero, circulated on social meeting. She inspired TV programs filmed in Georgia, including an episode of the Atlanta-based *Teenage Bounty Hunter* about a young Black woman decapitating Confederate statues (chapter 3), as well as the character of Angela Abar/Sister Night in *Watchmen*, and Anissa Pierce/Thunder in *Black Lightning*. (See Figure 5.1.)

When those shows were made, Confederate memorials were thus on the minds of many, including TV producers. Black superhero shows shot in Georgia engage the Confederate monuments in their locations as absent presences, represented indirectly, with symbolic stand-ins. But my final example, *Black Lightning*, directly confronts the monuments in its filming location, featuring a Black woman superhero who destroys a Confederate memorial in Atlanta.

Watchmen's first episode recreates the 1921 Greenwood Massacre, in which white racists attacked the prosperous Black neighborhood of Tulsa, murdering over 300 people, looting, and burning businesses, churches, schools, and homes.[2] When it first aired in October of 2019, the episode sparked extensive media discussions of racial violence and trauma in US history. The episode re-emerged in public discourse after the May 25, 2020, police murder of George Floyd, and again 15 days later when Donald Trump announced a rally in Tulsa, within walking distance of Greenwood. As if trolling Black people there and everywhere else for the pleasure of his white nationalist base, Trump first scheduled the rally for Juneteenth, but ultimately rescheduled it in response to public outrage. By contrast, to celebrate Juneteenth HBO made *Watchmen* available for free during the President's rally weekend. At the same time, Washington, DC, painted "Black Lives Matter" in large block letters in front of the White House, while other localities followed suit, including Tulsa's Greenwood neighborhood. These BLM paintings appeared to mimic the yellow color and typeface of *Watchmen*'s distinctive title sequences, leading *Watchmen* creator Damon Lindelof to suggest that the DC mural could open one of the show's episodes.[3] In these ways, the show spilled over into the real world, even as it obscured the race politics of the places in Georgia where it was filmed.

Watchmen centers on the intergenerational origin stories of Will Reeves as Hooded Justice (Jovan Adepo/Louis Gossett Jr.) and his granddaughter Angela Abar as Sister Night (Regina King), two masked Black heroes who fight the Klan. The program "imagines a redemptive narrative for superhero origins," according to Rebecca A. Wanzo, "both by writing a black man into the origin story and by making state-ignored (and state-generated) white supremacy the enemy."[4] Reflecting critically on our present reality, the series projects an alternate world where the police and the Klan are effectively the same institution. The Order of the Cyclops is depicted running the New York Police Department in the 1930s, while the Seventh Kavalry (sic) has infiltrated present-day police and government from Tulsa to DC.

In the opening scene of *Watchmen*'s first episode, depicting the Greenwood Massacre, a young Will Reeves sits in a Black-owned movie theatre watching a silent film of his heroic namesake, the Black sheriff Bass Reeves, at the moment Klansmen attack. The action next shifts to the present day, and we see a Black-cast production of the musical *Oklahoma* in the same venue, now rebuilt. The setting is inspired by the actual theater named the Williams Dreamland Theater in Greenwood, owned by John and Loula Williams, and destroyed by white rioters. Interior scenes were filmed in Macon's Douglas Theater, located blocks away from two Confederate monuments, one representing a giant generic Confederate soldier holding a rifle atop a tall pedestal and shaft, and the other a "Monument to the Women of the Confederacy" in the form of a large obelisk flanked by sculptures of two white mothers. The more extensive exterior scenes representing airplanes dropping firebombs, white looting, the murder of Black people by Klansmen, and the destruction of the theater, were filmed a few blocks from another large memorial representing a Confederate soldier with a gun, in the historic white commercial district of Cedartown, Georgia.

Cedartown is about an hour northwest of Atlanta, and home to a larger-than-the-national-average percentage of Black and Latinx people living below the poverty line.[5] Its current residents work for Walmart, corporate call centers, and office furniture manufacturers, but historically Cedartown was a segregated cotton mill town. Before the passage of the 1964 Civil Rights Act, Black workers were excluded from the mills and hence also from company housing.[6] Cedartown is named for a regional fort that was used as an internment camp for Cherokee people along the "Trail of Tears." The Georgia location's origins in anti-Black, anti-Indigenous racism thus anticipates the representation of settler white nationalism in *Watchmen*'s Oklahoma setting.

Cedartown's white leaders have resisted efforts to remove the marble "pedestal-shaft-soldier" memorial to the Confederacy located between the town's two courthouse buildings.[7] Complementing monuments to "great men" like Robert E. Lee, such statues depicting individual citizen soldiers are common in the region, where they were inexpensively mass produced and often the only piece of public art in many small towns. The basic statue of a soldier resting on a rifle while on watch depicts the Confederacy as defensive rather than aggressive. It also promotes a populist view of Confederate soldiers, balancing their commitment to a larger cause with their individual integrity as white men. Such statues symbolize and enforce race and class inequality in the town's past and present by putting everyday white supremacy on a superhero pedestal.[8]

Confederate memorials were mostly erected in two waves: around the end of the nineteenth century, under Jim Crow and as complements to "lost cause" narratives about the Civil War; and during the late 1950s and early 1960s, in reaction to the African American civil rights movement. According to Del Upton, they represented white supremacist historical revisionism. Upton argues that such memorials suggest that "those who fought for the white cause in the Civil War fought for 'their' state in a 'second war for independence.'" They also intimate that the Confederacy is an important part of the south's "heritage" and hence its monuments are sacred, the War was not about slavery, and that "soldiering is inherently honorable" regardless of the cause.[9] Their placement in front of courthouses and city halls symbolize white supremacist state power as a means of terrorizing Black people. In addition to its courthouse monument, Cedartown's Greenwood Cemetery includes the graves of almost 60 Confederate soldiers. Because, as Upton writes, "the very evidence of defeat provided by the graves suggested martyrdom in the cause of righteousness," cemeteries have often been chosen as homes for Confederate memorials.[10] Surrounded by hundreds of Confederate graves and memorials, it would be hard to make a TV series in Cedartown without noticing its many monuments to white supremacy.

Watchmen's other Georgia locations are also surrounded by Confederate memorials. The episode 3 scene where FBI agent Laurie Blake (Jean Smart) talks to Dr. Manhattan in a big blue phonebooth was filmed in Decatur's East Court Square, steps away from the Court House with its obelisk monument to the Confederacy, which would likely have been visible to cast and crew while shooting. Dispensing with personifications of the Confederate foot soldier, the obelisk can be read as an extreme expression of abstraction, the sculptural depiction of revisionary claims that the lost cause wasn't about slavery but valor and principle. *Watchmen* returns to East Court Square and its nearby obelisk in the series finale.

Sherriff Crawford's funeral in episode 3 was filmed in two Confederate cemeteries. Exterior scenes were shot at Atlanta's historic Oakland Cemetery, which includes a Confederate Memorial Ground where approximately 7,000 soldiers are interred, as well as two large memorials to the Confederate dead.[11] Interior scenes were filmed in the Decatur Cemetery, which includes numerous Confederate graves, as well as a Confederate memorial cross, erected by the United Daughters of the Confederacy in 1984, during the Reagan administration.[12] Like the Clinton-era monument at *Doom Patrol*'s Lawrenceville location, this

Watchmen location reminds us that a number of Confederate memorials were erected in living memory for many TV producers and viewers. Episode 6 of *Watchmen* is set in New York but filmed in Macon. The scenes where Reeves discovers a Klan plot to pit Black film audiences against each other were filmed in Macon's Hargray Capitol Theater, steps from the massive Confederate soldier statue and the Confederate mothers' memorial mentioned earlier. The scene of a Klansman burning a Jewish deli in that episode was shot nearby, on the same street. The town of Newnan, the location for the Hoboken carnival in episode 5, includes a granite memorial to William Thomas Overby, the "Nathan Hale of the Confederacy," erected in front of the local county courthouse in 1952, as well as one of the mass-produced statues of a generic Confederate soldier originating in 1868.[13] And in episodes 7 and 8, US-occupied Saigon was played by Griffin, a town with a Confederate soldier memorial and the Stonewall Confederate Cemetery. Finally, many other scenes were filmed at Atlanta Metro Studios. As a local reporter recently noted, the State Capital is home to numerous anti-Indigenous monuments and "a mother lode of Civil War and segregationist artifacts."[14] Confederate monuments are increasingly objects of vandalism, and Black people and their allies have agitated for their removal, yet Georgia's white ruling elite remain committed to preserving them as contemporary reminders that they are in charge.[15]

Although as far as I can tell none of the memorials I've described are visible in the show, in many ways *Watchmen* is in implicit dialogue with the Confederate monuments that dot the map of its Georgia locations. The program reframes the figure of the masked vigilante from comic books as a kind of memorial to the Confederacy and the white nationalism it embodies. Alan Moore, author of the *Watchmen* graphic novel, has argued that with their hoods and capes, Klansmen were models for comic book heroes, while historian Chris Gavaler suggests that *Birth of a Nation* helped inspire golden age comics.[16] The TV show draws on those histories with its depiction of the Order of the Cyclops and the masked members of the contemporary Seventh Kavalry. As part of a test to determine if a suspect is a white supremacist, Agent Looking Glass (Tim Blake Nelson) monitors their vital signs while they watch a slide show of monuments to American whiteness—fields of wheat, a milk advertisement, a man on the moon, a cowboy with a lasso, a Confederate flag, Mount Rushmore (with the addition of Richard Nixon), *American Gothic*, and photos of George Armstrong Custer and Klan rallies. Recalling Macon's statue of a larger-than-life Confederate soldier, *Watchmen* visually links settler colonial white supremacy to an aesthetics of white male monumentalism.

In the premiere episode, members of the Seventh Kavalry live in a trailer park called "Nixonville" decorated with a large statue of the President atop a pedestal in the style of a Confederate memorial, thus reminding us of Nixon's law and order, "silent majority" racism. Episode 3 prominently features a CGI-modified image of the Washington Monument, which was partly built by slaves to honor a slaveholder in the form of a giant obelisk, thus anticipating many Confederate monuments. As revealed in the final episode, Klansman and US Senator Keene dreams of stealing the massive powers of Dr. Manhattan in order to enforce white global domination. The avatar of such aspirations for white nationalist world domination is the Order of the Cyclops, named for the *Odyssey*'s giant one-eyed monster. In these ways, *Watchmen* suggests that the white supremacist monumentalism associated with the South has characterized US nationalism as such.

Watchmen's representation of white supremacy and the Vietnam War is more problematic. In its alternate history, Vietnam lost the war and is now a US state. In a *Washington Post* editorial titled "How 'Watchmen's' misunderstanding of Vietnam undercuts its vision of racism," Viet Thanh Nguyen faults the show for failing to represent US imperialism's "entwinement with white supremacy."[17] *Watchmen*'s depiction of Vietnam is underdeveloped and confused, but in episode 6 it does raise questions about the relationship between white supremacy and US imperialism. Recovering from an overdose of her grandfather Will Reeve's "Nostalgia" prescription, a drug that enables users to relive the past, Angela Abar vividly recalls her own childhood. In her memory, we see 10-year-old Angela as she walks through a Saigon street fair celebrating "VVN Day," a holiday commemorating the US victory over Vietnam. She stops to watch a puppet show depicting a giant Dr. Manhattan shooting flames from his hands at Viet Cong soldiers, intercut with images from the Greenwood Massacre of a Klansman threatening Black firemen with a torch. Dr. Manhattan's Vietnamese puppeteer gestures toward another man on a bike, who picks up a backpack and rides through the fair, his transit intercut with the image of another Klansman from the massacre scene. Finally, the bike rider leaps onto a military jeep shouting "Death to the Invaders" and detonates a bomb.

In a subsequent memory, young Angela is questioned by Saigon police, who ask her to identify the puppet master after they remove a hood from his head. As the police pull the man out of their car, the scene cuts to an image of Angela's grandfather, Hooded Justice, the anti-Klan vigilante, putting on his own hood, and then back to the scene in Saigon, where the police re-hood the Vietnamese bomber, who is then led away and shot off camera. Do these scenes suggest a moral equivalence between a Vietnamese suicide bomber and the Klan? Or between US white supremacy at home and abroad? The program seems to strive for a kind of modernist complexity or political balance that muddles sharp antagonisms. By shooting scenes of Saigon amid Confederate monuments in Georgia, however, *Watchmen* implicitly connects US imperialism and white supremacy.[18]

Yet by using locations in Georgia as stand-ins for Saigon, New York, and Oklahoma, *Watchmen* obscures its complicity in the conditions of white supremacy supporting its production in the state. While there are significant differences between the Greenwood Massacre with its violent destruction of Black lives and wealth on the one hand, and the upward redistribution of wealth represented by Georgia's tax credit program on the other, the two remain connected. Scenes of white characters looting Black neighborhoods in *Watchmen* suggest comparison to the less direct and more diffuse forms of violence represented by state incentive programs. As I argued in chapter 1, the promised employment benefits are limited and often concentrated in low-wage jobs, and in the case of *Watchmen*, the employment of extras is the exception that proves the rule. The show hired many local Black and white extras for the show's recreation of the Greenwood Massacre, and the job descriptions suggest one definition of TV racial capitalism in place: low-wage jobs of limited duration requiring Black people to recreate traumatic and traumatizing histories of racial violence. Because the industry depends on the upward redistribution of wealth from poor people to Hollywood, programs such as *Watchmen* benefit from tax subsidies at the expense of Black lives. Hence the state's lopsided budget priorities—incentives for film and TV production while starving social welfare—result from long histories of white supremacy and anti-Black racism in both Georgia and Los Angeles.

While in its content *Watchmen* opposes racism, in its mode of production it participates in and benefits from the racial capitalism that efforts to preserve Confederate memorials enforce. Moreover, by filming in Georgia and feeding into false narratives promoting the trickle-down benefits of corporate incentives, *Watchmen* further legitimates a status quo of racial inequality in the state. Acclaimed for its anti-racist narrative, *Watchmen* was subsidized by assets that a white nationalist state government stripped from poor people of color, which I would argue is less an irony than a mode of denial. Many appreciative accounts have been written of the show's groundbreaking narrative, yet its textual critique of racism distracts from the inequities surrounding its production. This is to say that the program's anti-racist content indirectly references yet ultimately displaces from view and critical reflection the intersecting race and class inequalities in their filming locations. By helping to deny its participation in the upward redistribution of wealth from poor Georgians to Hollywood, the anti-racism of *Watchmen* also helps encourage other shows to film there. One sign of this denial's effectiveness is the fact that, while regional news stories and local boosters celebrate Georgia's media incentive program, they rarely draw connections between what's on screen and the places where it was made.[19]

Two other Black superhero shows filmed in Georgia also address Confederate monuments in displaced forms, *Raising Dion* (2019–2020) and *Naomi* (2022), origin stories for young Black characters learning to use their developing powers. The first focuses on a widowed Black mother in Atlanta named Nicole, raising her young son as he begins to manifest superpowers in a world hostile to Black boys and men. Among other things, *Raising Dion* is about Black parenting, with Dion's conflicts at school and with friends serving as teachable moments. In a nod to *Watchmen*, for example, an episode called "Watch Man" suggests that Dion is punished at school because of super-powering while Black: when a white bully steals his watch, Dion telekinetically slams him into a wall, but only Dion is expelled. This sparks a talk from Nicole about how people will be afraid of him because he's Black (S1E3). Drawing on superhero figures like Professor X from *The X-Men*, Dion's best friend and ally, Esperanza, uses a wheelchair. When Dion employs his telekinetic power to make her walk, she angrily says he needs to respect her boundaries, and his mother uses the situation to educate him about consent (S1E9). Via stories about parenting a young Black superhero, *Raising Dion* thus stages critical reflections on race and disability. Many scenes in the show were filmed within blocks of the Decatur courthouse and its Confederate monument. While not represented in *Raising Dion*, I would argue that the Confederate memorials in its location are indirectly suggested by the show's supervillain, a large white, anthropomorphic creature called "The Storm." As Dion gains greater control over his powers, the monumental white Storm becomes his central antagonist, but he ultimately triumphs over the white monster. In their climactic battle, the Storm towers over Dion, shooting lightning bolts from its hands at Dion, who counters with lightning of his own. With his mother's help, he vanquishes the creature by piercing its heart with a lightning bolt. The vanquished supervillain can be read as a speculative revisioning of the white supremacist statues where *Raising Dion* was filmed, a sort of sublation that simultaneously preserves and cancels local Confederate memorials.

Naomi, created by Ava DuVernay, also uses superhero conventions to investigate race and gender in a coming-of-age story. Set in the Pacific Northwest but filmed in Georgia, the title character (Kaci Walfall) is a Black, bisexual high school student and a Superman super fan who begins to manifest powers herself. Naomi is adopted, and its ultimately revealed she's from another planet: hence, her origin story follows Superman's arc. In the premiere, Naomi skateboards to school through the historic downtown square of Decatur that was also used in *Watchmen*. A friend texts her that there is some sort of "stunt" involving Superman in the square, so she skateboards back but faints before she can discover what has happened. Friends tell her Superman landed in the square and fought with a villain. Naomi collects cell phone video of the event and pieces it together, revealing that just before flying away, Superman hovered above one of the square's distinctive blue towers, making it appear as though he is on a pedestal, like the filming location's nearby Confederate soldier memorial. *Naomi* also sublates the Confederate monuments in its midst. (See Figure 5.2.)

The preceding programs hide contradictions. Examples of DEI TV, they present anti-racist narratives that obscure their complicity in

Figure 5.2: Superman hovering over a pedestal like a Confederate memorial in *Naomi*.

structural inequality. By contrast, *Black Lightning* foregrounds contradictions. The show is forthrightly set where it was shot, in Black Atlanta, and focuses on the Black superhero of the title, Jefferson Pierce (Cress Williams), the idealistic principal of Garfield High School. His daughter, Anissa (Nafessa Williams), is a teacher there, and his youngest, Jennifer (China Anne McClain), is a student. Like *Raising Dion*, *Black Lightning* is partly about Black parenting and family life as the daughters begin to manifest powers like their father's. In season 1, episode 6, Anissa attends a university protest of a Confederate monument, a generic general holding a rolled-up document (a prop and not an actual statue). A group of young people surround the statue and spray it with red paint from toy guns. They are arrested, and Anissa's father bails her out. "That statue, dad, is an insult to Black people." "You broke the law, Anissa, that gives the police an excuse to hurt you, and on top of that you used guns. . . . All it takes is for one cop to see your color instead of your humanity and decide better dead than sorry. Look, you're a black woman. You don't have the luxury of being naive." Here and elsewhere, *Black Lightning* dramatizes generational differences over Black political strategies. Like the white nationalists at the Unite the Right rally in Charlottesville, on the show white men subsequently surround the statue shouting "you will not replace us!" They are met by counter-protesters, and a white nationalist drives into them, killing a 19-year-old student. Later, dressed in

Figure 5.3: Thunder destroys a Confederate monument, *Black Lightning*.

costume as Thunder for the first time, Anissa goes back to the memorial, which is surrounded by flowers, candles, and mourners, and stamps her foot on the ground, making the ground ripple and the statue explode. (See Figure 5.3.)

The destruction of the Confederate monument becomes her superhero origin story. Whereas other programs cut the monuments out of the frame, *Black Lightning* centers one. Moreover, while other shows silently divert funds from education and other social spending, or even project negative representations of social welfare, as in horror shows about foster care (chapter 4), *Black Lightning* represents the values and challenges of public education, making the character of Black Lightning a principal and Thunder a teacher at an underfunded Black public school.

Alien and Migrant Detention in *Roswell, New Mexico*

The town of Las Vegas, New Mexico, has often been employed as a location for film and TV westerns, starting with dozens of silent films starring cowboy star Tom Mix. More recently, Las Vegas appears in the film and TV westerns *No Country for Old Men* (2007) and *Longmire* (2012–2017). The historic Las Vegas commercial district around the train station was also used for the hybrid horror western TV show

Midnight, Texas (chapter 4), while the competing commercial district around the old plaza appears in the science fiction westerns *Outer Range* and *Roswell, New Mexico*. The last two were both partly filmed on the same block of Bridge Street, which ends at one corner of the historic town plaza. The location is sedimented with histories of settlement and empire. The plaza was first built by Mexican settlers in 1835, on a Mexican land grant. Standing on top of an adobe building in the plaza eleven years later, General Stephen W. Kearny delivered a public proclamation claiming New Mexico for the United States as part of the US/Mexico War. Walking through the plaza today, cast and crew for *Outer Range* and *Roswell, New Mexico* might encounter the proclamation inscribed on a bronze historical marker there, which reads in part "Mr. Alcalde, and people of New Mexico: I have come amongst you by the orders of my government, to take possession of your country, and extend over it the law of the United States. . . . We come amongst you as friends—not as enemies; as protectors not as conquerors. We come among you for your benefit—not for your injury."

The bronze marker implicitly memorializes the violence of the war, and the dispossession of Indigenous and Mexican people, making it a southwestern analog to Confederate memorials. Moreover, although the two local contexts are very different—the US military and railroad invasion of Las Vegas in the nineteenth century, and a contemporary film and TV industry often aligned with policing and prisons in the twenty-first—plaza visitors reading that inscription might rightly be reminded of Hollywood's claim that it benefits people in New Mexico. Left off the marker, and without one anywhere of its own, is the 1847 Battle of Las Vegas, where the US army crushed a revolt by Mexican and Pueblo people. Direct references to the US/Mexico War are similarly absent from the many films and TV shows made in New Mexico. This is no doubt partly because of the relative historical amnesia about the war, but it might also be because images of invaders suggest uncomfortable comparisons to Hollywood.

Roswell, New Mexico (2019–2022) was produced by a diverse group of actors, directors, and writers.[20] Given its use of conventions for representing aliens to engage the recent history of anti-immigrant rhetoric and policies in the United States, it is not surprising that the show's writers included three Los Angeles-based Chicanas: Leah Longoria, an

Amherst film studies graduate originally from Austin, Texas who also wrote for *Jane the Virgin* (2014–2019); Carolina Rivera, who also served as a co-executive producer for the show and as a writer for *Jane the Virgin*; and Ariana Quiñónez, a mariachi musician from Bakersfield who graduated from Loyola Marymount University in Los Angeles and who was a staff writer for *Home Economics* (2021).[21] This helps explain the show's sharp, critical dialogue about race and immigration and its empathetic representations of undocumented migrants living under the threat of deportation and family separation.

Over the course of its four seasons, *Roswell, New Mexico* focuses on two "alien" families: the Mexican Ortechos, including Liz, her sister Rosa, and their undocumented father Arturo, who runs an extraterrestrial themed Mexican restaurant called "The Crash Down"; and orphaned space alien siblings Max Evans (the local police deputy), Isobel Evans, and Michael Guerin. (They have different last names because the first two were adopted by human parents while Michael bounced between foster homes.) Both families work hard to avoid detection, the Ortechos because of the prospect of deportation, the aliens for fear of becoming subjects of government experiments. The intertwining of the two families enables critical commentary on white nativism, as well as the projection of its antithesis in the love, kinship, and alliances among different kinds of humans and aliens.

Roswell, New Mexico dramatizes the risks of deportation when seeking medical care, as well as the health risks of ICE detention. Visiting a friend at the hospital with her father, Liz overhears an ICE agent arguing with a doctor. The agent looks over his left shoulder and makes eye contact with Liz. When agents arrest a young Mexican man and his grandmother, the camera cuts again to Liz, looking worried and wide eyed. She scans the waiting room, frantically looking for her father. As she grabs his hand to lead him away, the first ICE agent approaches them and asks for IDs. The shot is a low-angle one from behind the pair's heads and with the agent's shaved white head in the center, looming large over the undocumented father and his daughter. This is followed by a reverse shot of Arturo and Liz, who stare back with big eyes and exchange worried glances. Over Liz's protest that he has applied for a green card, she is his sponsor, and that he shouldn't be detained anyway because he's diabetic, the ICE agents take Arturo away. The scene's focus on eyes and glances suggests the experience of being under ICE's gaze.

With Arturo's ICE detention, the show further foregrounds the recent history of migrant deaths in detention. At the detention center, Liz tries to deliver her father's medicine, but a guard says he must rely on the infirmary inside. Liz sarcastically asks, "did Nebane Abienwi visit an infirmary before he died of a brain bleed in your custody," referring to the 2019 death of a Cameroonian asylum seeker who died in ICE custody in San Diego.[22] She continues, "What about Johana Medina León, she was twenty-five years old," referring to a trans asylum seeker from El Salvador who had been detained at a New Mexican detention center in Otero County and then subsequently died in an El Paso hospital after being denied medical care. Afterwards, ACLU New Mexico met with a dozen gay men and transgender women at the detention center "who recounted harassment, denials of medication and the use of solitary confinement to dissuade people from issuing complaints."[23]

In opposition to white vigilantes, racist police, and ICE agents, *Roswell, New Mexico* represents alternative ways of living and being in the world. The show depicts diverse romantic couples: heterosexual and queer, Black and Latinx couples, white and Latinx, and alien and human. Season 1, episode 2 ends with two sex scenes, parallel edited together: Max and his partner, Jenna Cameron, and the alien Michael Guerin and the closeted soldier Alex Manes. In addition to racists, the characters in *Roswell, New Mexico* also square off against homophobes, like Alex's soldier father, who hates aliens, immigrants, and queer people, including his son. By contrast, the alien deputy Max Evans is a self-described feminist, and when Arturo is attacked by a white racist, Max shields him from deportation (S1E2). And while she is ultimately replaced by Taylor, at the start of the series the local Sherriff is a feminist Mexican woman critical of patriarchy, Michelle Valenti (Rosa Arredondo). Such utopian representations of the police draw into critical relief the limitations of actual police institutions but risk romanticizing individual officers in ways that undermine audience appreciation for the structural nature of police violence. Handsome and enlightened Sheriff Evans, after all, is one of the show's central romantic leads, an impossibly "good cop" whose existence blunts *Roswell, New Mexico*'s critical edge by making it DEI TV about DEI police.

Indeed, the show's settings, cinematography, and lighting combine to romanticize prison architecture in New Mexico. Scenes of the town of Roswell, including the Crash Down Café and the downtown plaza, were filmed in Las Vegas, New Mexico, but many other scenes were shot on Santa Fe Studios soundstages and the land surrounding them. Throughout the series, for dramatic visual effects, *Roswell*'s producers often shot on a desert plain, with buildings in the distance, twinkling with lights at night, between the action in the foreground and mountains in the distance. In many such scenes, the prison complex across the street from Santa Fe Studios serves as backdrops to interactions among characters, especially romantic ones. A local carnival includes a makeshift drive-in movie theater showing the campy film *Queen of Outer Space* (1958), which is projected on a screen mounted on large, stacked storage containers. Prison lights are visible in the background, alongside many other twinkling lights in the romantically shot scenes of couples at the carnival film screening (S3E3). The location is also used for a subsequent carnival screening, where the white alien Isobel confesses her love for her Black human girlfriend Anatsa in the rain (S4E1). In these scenes, prison is effectively aestheticized as romantic mood lighting.

Similarly, in a later episode Michael and Alex are trapped in a "pocket dimension," set in the same prison-adjacent location but filmed through a pale purple filter, as if at twilight. The alien and human lovers reminisce about their first date. Alex says, "All of these restless nights, I just sat looking at this blank screen, imagining that this nightmare was just a movie about star-crossed lovers. I'd sit here dreaming about getting home to you and I could picture you perfectly on the other side doing the same." At the end of the scene, he proposes, and after Michael says yes, they kiss and fall back into the bed of an old truck, the pale purple prison buildings visible behind them (S4E11).[24] The outdoor film screen location that *Roswell, New Mexico* used for this scene, which is across the road from a prison, is a concrete example of what I theorized in chapter 2 as Hollywood's adjacency to carceral institutions. Publicly subsidized, the studio and the New Mexico Corrections Department share funding models, for example. Both Santa Fe Studios and the surrounding prisons depend upon a network of publicly funded roads connecting them. These commonalities encourage TV representations of diverse heroes of law enforcement that obscure structural inequalities in the criminal justice system, not to mention how state-subsidized shows such as *Roswell, New Mexico* divert funds from social spending and education and low-income housing as alternatives to policing and prison. On the landing page for its website, Santa Fe Studios advertises a "30% NM Cash Rebate" with a promotional image of the *Roswell, New Mexico* cast. By employing the show to promote state subsidies, the studio also advertises state power, one branch of which is the local Department of Corrections.

What are we to make of a show with a narrative critical of ICE that nonetheless romanticizes prison? It is partly a result of the regional differences, including class differences, between Los Angeles, where series writers are located, and New Mexico, where their scripts are transformed into shows in concrete, local settings. From such a distance, it can be harder for Latinx creatives to see how their critical perspective on migrant detention works as a form of DEI TV, in which pro-immigrant representations effectively legitimate (by obscuring) how, in its production, *Roswell, New Mexico* participates in the police and prison televisual complex. Although *Roswell, New Mexico* includes more sharply critical content than *Devious Maids*, the two shows nonetheless represent Latinx racial capitalism in place.

Outer Range also combines narratives about aliens and the police, but with a crucial focus on settler colonialism. It centers on a Wyoming cattle ranching family named the Abbotts and their neighboring rivals the Tillerson family, who have made legal claims to a large tract of the Abbotts' pastureland. The first episode is titled "The Void," in reference to the large cosmic black hole in the ground that Royal Abbott (Josh Brolin) discovers on his ranch. The portal is a swirling dark mass, and the episode begins with Royal throwing a body into it. Later we see the corpse, dressed in western wear, sink in the portal, and disappear. The body is Trevor Tillerson, who died in a bar fight with Royal's sons Rhett and Perry, and the family, including their mother Cecilia (Lili Taylor), work overtime to hide the crime. The black hole seems to be a time travelling portal. Royal confesses to Rhett that he is from the nineteenth century and used the portal to travel to 1968 after he accidently shot his father while hunting. Covering up one death in the past, Abbott covers up another in the present, suggesting the continuity of settler violence across time.

But the black hole remains mysterious, and the show raises more questions than it answers. Other than hints of time travel, *Outer Range*'s creator and showrunner Brian Watkins seems to purposely frustrate efforts to find meaning in the void, and, by extension, attempts like this one to investigate the connections between New Mexican locations and fictional settings. Even so, one actor in the show, Tamara Podemski, provides revealing commentary on that topic. Podemski's mother was Anishinaabe from the Muscowpetung band of First Nations people in Saskatchewan, and she has enjoyed a distinguished career in Canadian and US theater, film, and TV. Her sisters, Jennifer and Sarah Podemski, are also accomplished film and TV actors; together they co-starred in *Reservation Dogs*. Tamara Podemski has played police investigators several times: in the films *Never Saw It Coming* (dir. Gail Harvey, 2018) and *Guest of Honor* (dir. Atom Egoyan, 2019) and on the TV show *Coroner* (CBC, 2019). As actor Gary Farmer (Cayuga Nation and Wolf Clan of the Haudenosaunee/Iroquois) said when he interviewed Podemski, "You seem to play a lot of cops."[25] On *Outer Range*, she plays acting Sheriff Joy Hawk, a queer Indigenous woman who investigates the Abbotts.

In interviews about the show, Podemski explicitly addresses the challenges Indigenous women and women of color negotiate working on DEI TV shows. From the start, the actor's aim was to insure that "Joy wouldn't get *swallowed up* by this toxic masculinity and this heavy kind of romantic western" (emphasis added), as if to suggest that the devouring portal is the maw of manifest destiny. To prevent Sheriff Hawk from being drawn into that vortex, Podemski knew she would need to make the character compelling in her own right, but that was difficult given the script.

> It was very clear that this was a character who was given every diversity check all pushed in and rammed into one role. And not only that, it wasn't a realistic role. There's nowhere in reality, in America, where this woman exists, at the time, you know, two years ago when I first read it. And I think my research showed very few women of color that were sheriffs and even those that were, were near more metropolitan areas. So, to take her and set it in such a rural place, and make it even more challenging, to make her an out gay woman running for office, and Native American, I found the burden of that was much greater because I needed

to ground her in reality. And I needed to make it believable and that, I think, is where I spent most of my time figuring out how to not let her just . . . *fill the diversity hole* in the show, but to have her be a fully rounded three-dimensional character who exists in that world, and not to just be a foil, but to actually be an oppositional force (emphasis added).²⁶

Podemski's performance is remarkable, and, as I argue below, she succeeds in turning the character into "an oppositional force." But her expression of fear that the show might swallow up her character, and that she might sink into the "diversity hole," suggests one interpretation of the big hole in *Outer Range* as a metaphor for the pitfalls of working in DEI TV.

From Podemski's perspective, the challenges she faces as an actor are mirrored in her character. At the start of the series, the previous sheriff has recently died, and Hawk is promoted to acting sheriff while an election campaign begins. She courts the support of voters, including by attending the Abbotts' Catholic church with her wife, Martha (Morningstar Angeline, Navajo, Chippewa, Blackfeet, Shoshone, and Latinx), and daughter Rose (Ofelia Garcia), where a deacon welcomes them with a homophobic prayer about how marriage is between a man and a woman. Angered, Martha wants to leave, but Joy persuades her to stay. Podemski interprets the scene as emblematic of struggles over representation for Indigenous performers: "For all of us that have been fighting for our voices to be heard, fighting for representation, fighting to see our authentic lives represented on the screen, it's important to understand that the fight comes at a cost. Even on a personal level, we know how many things we have to compromise to push something forward a little bit." This is an elegant description of a "war of position" within DEI TV waged by Indigenous actors and actors of color.

Podemski's view of her role suggests an Indigenous perspective on the hole and its appearance to white ranchers that further implicates the process of TV production.

> The interesting thing about the hole, which Joy never sees, is that there are Indigenous stories in Montana and Wyoming about these time portals on reservations. It was asked of (series creator and co-showrunner) Brian Watkins, and he'd never heard of these portal stories on reservations.

Figure 5.4: Sherriff Joy Hawk, *Outer Range*.

> This was just what his own imagination dreamed up. That's why I resist calling it sci-fi because to me these are stories I've heard. These multi-dimensional realms and universes are very common in some of our Indigenous stories. I don't think Joy is as freaked out by the crazy stories. What's weird to her are how these repressed, cut-off, or unspiritual people interact with the hole. That's the thing that's so very bizarre.²⁷

This suggests that one way the character represents an "oppositional force" is in the form of an Indigenous epistemology at odds with the rancher's toxic white masculinity, or, as she told an interviewer, "If you want to look at *Outer Range* as an example of colonial insanity, then Joy really offers an alternate approach to the madness that is going on."²⁸

Podemski's interpretation is supported by the season 1 finale, where various characters inexplicably encounter a huge herd of stampeding buffalo. Whereas others encounter the buffalo as a terrifying danger, threating to trample them, or run their cars off the road, Sherriff Hawk appears exhilarated by them. The cold opening of the episode follows the character as she hikes through a forest of aspens and pine and hears a loud rumble. The sound gets louder as she approaches a clearing, and the camera closes in on Hawk's face as she smiles, almost laughing. The next shot is from behind her as she looks at the herd running across the valley floor. The camera ultimately returns to her face as she smiles again. Later, in Hawk's final appearance in the episode, she looks down into a valley from a mesa top and sees the herd again, but this time it is across the river from an Indigenous settlement of teepees. As she looks

Figure 5.5: The buffalo return, *Outer Range*.

on, several Indigenous figures hunt the buffalo, and the scene again turns to a closeup and the expression on the character's face is one of awe, as with trembling lips she silently mouths the word "oh."

When asked by the *Hollywood Reporter* what her character thinks when seeing the buffalo, she answers that it "really is all happening in real time," and that "her truest reaction to what is going on is the deep knowing that this is all real. And I wanted to leave her in that space of the heartache, of the longing, of the warning, of the weight of what it is."[29]

This insistence that the scene is real, happening in real time, reminds viewers that struggles over Indigenous land are historical but also contemporary and ongoing. By emphasizing the immediacy of such struggles, Podemski implicates TV production in the here and now of settler colonialism.

6

Black Atlanta TV

The Oprah Winfrey Network (OWN) show *Greenleaf* (2016–2020) is set at a Black megachurch in Memphis but filmed in Atlanta. In season 1, a Black police officer and church member shoots an unarmed Black teen, leading to local "Black Lives Matter" protests. The mayor responds by asking a skeptical Bishop James Greenleaf (Keith David) to deliver a "Back the Blue" sermon in exchange for a parcel of land owned by the city. Although the show does not make the connection, the scenario of state-subsidized land in exchange for supporting the police works like an allegory for how government media subsidies promote prisons and policing in Georgia and New Mexico. (See chapter 3.) Over opposition from the board of deacons and his wife, Lady Mae (Lynn Whitefield), Bishop Greenleaf goes ahead with the sermon, which ends with a group of about 20 uniformed police officers (Black and white, men and women) joining him on stage (S1E4). (See Figure 6.1.) "It's actually a pretty tense moment," writes reviewer Nichole Perkins. "We don't know if the officers are there as a sign of support or as a threat. Honestly, it might be both. Their rigid stances don't really say, 'We're here to protect and serve.' They look very intimidating, even in the house of the Lord."[1]

Indeed, as the police file onto the church stage, the choir sings "The Battle Hymn of the Republic," with its ominous reference to God's "terrible swift sword." The board of deacons walk out in protest, and the bishop ultimately rejects the mayor's offer of land, but the emotional weight of *Greenleaf* is behind the police officer, depicted as a sympathetic figure who deeply regrets his mistake. Viewers are encouraged to empathize with him because the moral center of the show, the bishop's daughter Pastor Grace Greenleaf (Merle Dandridge), takes up the officer's cause, and lends a sympathetic ear as he complains about being vilified by BLM protestors. Although the dead teen and the circumstances of his shooting are never represented, at the end of episode 9, his brother shoots the police officer in front of Grace, who cries and yells for help.

Figure 6.1: Bishop Greenleaf welcomes the police, on *Greenleaf*.

In a shot-reverse-shot sequence, the camera cuts from a closeup of the dying policeman to Grace's emotional face before the screen shifts to black. (See Figures 6.2 and 6.3.) In the aftermath of highly publicized police shootings of Black people near the time of its production, *Greenleaf* constructed a narrative of formal equivalence, in which a police officer and the unarmed teen he shoots are both innocent victims. This narrative arc, I would argue, is an expression of a police and prison televisual complex, or the set of mutually beneficial relationships between Hollywood and the criminal justice system. Such relationships help sustain TV racial capitalism in place, and their traces here, in a progressive Black show filmed in Atlanta, suggest the need to elaborate Cedric Robinson's theories to account for forms of Black TV racial capitalism in place whereby more privileged Black creatives benefit at the expense of poor Black and Latinx people.

The Atlanta-made TV shows analyzed in this chapter represent elite Black institutions and figures in religion, education, law, politics, business, and the media. *The Quad* (BET, 2017–2018) is set at a fictional HBCU. The main characters in *Ambitions* (OWN, 2019) are lawyers, Atlanta politicians, and CEOs. *The Oval* (BET, 2019–) is about a fictional first family. *Being Mary Jane* (OWN, 2013–2019) focuses on a Black TV news reporter. From the vantage of such characters, Atlanta-shot shows present complex, critical representations of anti-Black racism and sexism

Figure 6.2: A policeman shot on *Greenleaf*.

Figure 6.3: Grace Greenleaf looks on as the policeman dies.

in the MAGA era. But they focus their critical representations through wealthy Black characters who wear designer dresses, suits, jewelry, and handbags; own luxury cars (often with drivers); and live in gated homes and mansions filled with expensive furniture and art, grand bathrooms, and clothes-filled walk-in closets. These lavish homes are cleaned by uniformed maids, often Latinx women of uncertain legal status. As one reviewer of *Being Mary Jane* wrote about an episode where the title character and her boyfriend debate whether to vacation on a private island or in the Hamptons, "my wallet hates this TV show."[2] In the form of public subsidies for film and TV production, the upward redistribution of wealth from poor Black and Latinx people to more privileged Black media makers pays for depictions of Black wealth. Complementing the representation of Black capitalism, programs in this chapter also represent working-class Black and Latinx life from middle- and upper-class Black perspectives that seem to legitimate diverting public resources from social welfare to TV production. Another common thread among the different shows made in Georgia is a relative support for the police. I devote the most extensive attention to Terry Perry's public persona and his sensational White House melodrama *The Oval*, and to Mara Brock Akil's TV news drama *Being Mary Jane*, since both programs crystalize the contradictions of respectability politics and Black racial capitalism in the genre of Atlanta-based melodramas.

Black Institutions and Leaders on Atlanta-Made Melodramas

In *The Legend of the Black Mecca: Politics and Class in the Making of Modern Atlanta*, Maurice J. Hobson historicizes the representation of Atlanta as a "mecca" for Black people, tracing its origins to Reconstruction, when the presence of the Freedman's Bureau district office made the city an attractive destination for Black migrants. But in the final decades of the twentieth century Atlanta was called the "Black Mecca" because of Black educational institutions. As Hobson writes, it is home to "four black colleges, one black university, and one black seminary center. This cluster of institutions of higher education had no parallel in the rest of the United States or the world and played a critical role in establishing and maintaining Atlanta's thriving black upper and middle classes. Yet these institutions also created a nepotistic and exclusive black caste and class system that greatly influenced the city's politics."[3] Upper- and middle-class access to Black higher education, in other words, paved the way for political power in the form of elected office. Educated Black elites in government, Hobson argues, have often made alliances with white capitalists at the expense of working class and poor Black people. "A divide between the black elite and the black poor had always riven Atlanta's social fabric. Even after the city government shifted from white to black hands, its leaders pursued policies that benefited white and black elites to the exclusion of the vast majority of the black citizens who had brought them to power."[4] Hobson devotes a chapter to the history of the Atlanta Olympics, in which Black politicians and white corporations used public subsidies to build stadiums and other athletic infrastructure that displaced poor Black people from homes and communities. In such contexts, he concludes, discourses of "racial pride overcame attention to issues of class" as Black elites in Atlanta represented their own enrichment as progress for Black people as such.[5] Black TV melodramas made in Atlanta participate in these historic dynamics, paying for narratives centering the ethical and political perspectives of wealthy Black professionals and political leaders by redistributing wealth upward, from less to more privileged Black people. In contrast with what Robinson famously called "the Black radical tradition," the shows in this chapter represent Black TV racial capitalism in place.

Black TV melodramas made in Atlanta is a distinctly regional genre.[6] Such shows represent forms of worldmaking focused on Black institutions and Black leaders. They often use Black churches as important settings, for example. Similarly, Historically Black Colleges and Universities (HBCUs) are referenced to varying degrees in several shows, while *The Quad* is set at a fictional HBCU called "Georgia A&M," and depicts marching bands, football, fraternities, and Black intellectual life. Professor Grace Caldwell teaches a course on "Race and Media" where students read Frantz Fanon, bell hooks, and Stuart Hall (S2E2). Dr. Caldwell is played by Jasmine Guy, known for her role as a college student at a fictional HCBU in *The Cosby Show* spin-off *A Different World* (1987–1993), thus suggesting an intergenerational history and continuity in both Black higher education and in Black TV production.[7] Students are also pictured reading Octavia Butler's vampire novel *Fledgling* (S2E10); Jon Lewis's *March!*, a graphic novel about the civil rights movement which gives its name to the episode (S2E6); and Charles Johnson's novel *The Middle Passage*, which is also the episode's title (S2E7). Indeed, every episode is named for a different example of Black literature, including *The Invisible Man*, *Their Eyes Were Watching God*, and *The Color Purple*. Similarly, on *Being Mary Jane*, Atlanta TV reporter Mary Jane Paul (Gabriel Union) engages contemporary Black intellections. In one editorial, she references Michelle Alexander's *The New Jim Crow* (S1E8), while on a TV panel about anti-Black beauty norms she interviews Duke University Professor Mark Anthony Neal (S2E9).

Shows shot in Georgia also represent the world of Black politics. *Ambitions* dramatizes the rivalry between two former friends and Spellman classmates, one of whom is married to the fictional Atlanta Mayor Evan Lancaster. Meanwhile, *Being Mary Jane* includes an episode in which Paul reports on the dedication of a building named after Thurgood Marshall with real-life Atlanta Mayor Kasim Reed (2010–2018) and Georgia political figures such as John Lewis and Andrew Young (S4E14). Atlanta programs feature Black lawyers and judges, including Assistant US Attorney Amara Hughes and her lawyer husband Titus (*Ambitions*); corporate lawyer Sheldon (*Being Mary Jane*); divorce lawyer Andrea Barnes (*Sistas*); and Deputy D.A. Lila Standrich and Judge Charles M. Richardson (*The Quad*). Such programs also showcase Black CEOs and venture capitalists, including Beta New Electric CEO Marcin Barnes (*Ambitions*), Fortune 500 CEO Gary Marshall Borders (*Sistas*), and venture capitalist David Paulk (*Being Mary Jane*). Finally, *Being Mary Jane* and *Love Is_* (OWN, 2018) focus on the world of Black network TV producers who struggle with institutional racism and sexism.

Such programs present critical perspectives on issues faced by Black communities, such as anti-Black media bias (*Being Mary Jane, Love Is_*), sexual violence (*Greenleaf, The Quad, The Oval*), and gentrification (*Ambitions*). They also respond to topical issues of the MAGA era. When Black students on *The Quad* protest a proposed merger of Georgia A&M with a predominantly white institution, a group of white students stage their own protest, with "All Lives Matter" signs, the Confederate flag, and a Nazi-inspired alt right flag that, roughly a year before the episode screened, was flown by white nationalists during the infamous "Unite the Right" rally in Charlottesville, Virginia.[8] One character also wears a red hat resembling those worn by Donald Trump fans (S2E6). Similarly, in season 4 of *Being Mary Jane*, the TV reporter comes in conflict with the network's new alt right reporter Dani. Seemingly inspired by far-right media personality Tomi Lahren, Dani spouts casual racism in the workplace, leading Mary Jane to dramatically dress her down in front of the entire newsroom.

Both shows also represent police violence. Season 2, episode 4 of *The Quad* dramatizes the dangers of driving while Black when two white police officers wrongly arrest President Fletcher, pointing a gun at her and slamming her against the police car. In a particularly harrowing sequence at the police station, the arresting officers strap her to a table to draw blood, suspecting she's on drugs, as Fletcher protests. The episode ends with cross cutting scenes of Dr. Fletcher being interviewed by CNN contributor Roland Martin about her arrest, and scenes of a student, who had earlier in the season been violently arrested in his dorm room (S1E3), watching a TV interview with Gwen Carr, the mother of Eric Garner. Garner was killed in 2014 by a New York Police Department officer using a prohibited choke hold. Carr's interview is conducted by Felicia D. Henderson, *The Quad*'s creator and showrunner, as well as an Associate Professor in the Radio/TV/Film Department at Northwestern University.

> FLETCHER: The Policing of African Americans is a systemic issue this country has been struggling with since Black people were first enslaved here. And back then, we were only seen as property, so it was completely acceptable to beat, harm, kill us as methods of discipline.
> CARR: The day he was murdered, he was not selling cigarettes that day (as police alleged). He had just broken up a fight.
> FLETCHER: Exactly. Now how long must this go on? It has to stop.
> CARR: And I was just screaming at the TV, and I was saying, "Please, y'all, stop. Let him go." You could go around killing our children and just getting away, literally getting away with murder.
> FLETCHER: But if a 5'2", 120-pound Black woman in a Jag wearing a suit is a threat to the police, what chance does everyone else have?
> CARR: And there are so many mothers out there who suffer as I do, and if you listen to the stories, one story is worse than the other.

The show then cuts to footage from the *Daily News* website of Garner in a police chokehold, repeatedly saying "I can't breathe."

This painful, powerful episode of *The Quad* exemplifies Herman Gray's claim about Black TV shows and the screening of Black trauma.

Citing *Underground* (2016–17), *The Underground Railroad* (2021), *Atlanta*, *Lovecraft Country*, and *Watchmen* (all filmed in Georgia), among other shows, Gray argues that

> The subject position of the ideal viewer anchoring Black-cast and -themed historical dramas . . . makes visible—perhaps for the first time—a conception of Black trauma operating both on the screen and off. . . . The meaning of Blackness posited in these (shows) illustrate the multiple, interlocking, and dynamic conceptions of Blackness and the histories of trauma and pain that are traced and registered televisually. . . . With media coverage of the murders of Sandra Bland, George Floyd, and so many others, Black television viewers often respond with complex emotions to repeated televisual rehearsals and displays of violence perpetrated against Black folks and inflicted at the hands of police.

Gray concludes that in response, Black shows address Black viewers of police violence "through forms of identification and healing, care and intimacy."[9] As Christine Acham argues, "the rise of BLM (the Black Lives Matter movement) forced open a space on TV television and streaming industry for Black stories and storytellers," and hence Black viewers.[10] This kind of televisual address to Black witnessing of police violence is performed by the Eric Garner episode of *The Quad*, but also represented within the show, in the figure of the Black college student and victim of police violence who watches Henderson's interview with Carr.

The Garner episode, however, is followed by one that promotes a liberal, reformist perspective on the police. As Fletcher reads Alice Walker's *In Love and Trouble*, she's texted by her friend, Assistant DA Lila Standrich, who tells her to turn on Channel 3. The TV reveals Standrich at a news conference announcing a formal review of the arresting officer and the prospect of unnamed police "reforms." Fletcher texts back, "Justice was served today." Previously, she had claimed that anti-Black police violence was "systemic," suggesting the need for police abolition, yet the show ultimately settles on a reformist logic which, historically, has served to shore up and extended police violence.[11] *The Quad*, however, represents this outcome as a triumph of Black feminism.

From Homeless to the White House: Tyler Perry and *The Oval*

Scholarship about Tyler Perry has focused attention on the problematic content of his many films and television shows. While noting his work's appeal to Black audiences, Black feminist film and TV critics have criticized Perry's stereotypical representations, particularly of women; patriarchal respectability politics; and emphasis on individual acts of love and forgiveness in the face of structural inequality and violence.[12] In her study of melodrama in formal politics and popular culture, *Re-Imagining Black Women: A Critique of Post-Feminist and Post-Racial Melodrama in Culture and Politics*, Nikol G. Alexander-Floyd argues that Perry's films "center on and romanticize patriarchal family models, play on stereotypical images of black women and feature narratives of self-help or personal life makeovers that stand in sharp opposition to radical black feminism."[13] She suggests that Perry represents a combination of respectability politics and neoliberalism in which material social problems can be solved by individual responsibility and heterosexual coupling instead of political organizing or government social spending. Here I build on such research, including work that connects the content of Tyler Perry films and TV shows to the political economy of their production. Perry's production model merges a Christian belief system with capitalism alongside ideals of rugged individualism, efficiency, productivity, innovation, and frugality.[14] Journalistic accounts of and interviews with Perry celebrate his media empire in ways that indirectly reference its dramatic upward redistribution of wealth as well as his support for the police. In what follows, I use the keywords "Mass," "Flight," and "Ownership" to analyze Perry's production model and largely celebratory media coverage. The terms are also key to understanding the Black racial capitalism in place that he represents.

Mass. Articles about Perry and stories on his website are filled with metrics of his outsized success. Profiles detail his numerous honors and awards, the massive size of the 300-acre Tyler Perry Studios (reportedly the largest in the world), and his films' huge opening weekends and other box office records. *Forbes* estimates that his "Madea" franchise alone has made more than $660 million. Journalists also report record ratings for his TV shows. The season 3 finale of *The Haves and Have Nots* reached 3.71 million viewers, making it the network's most watched show ever. Season finales of *The Oval* have similarly attracted between 2.9 and 3.2 million viewers.

Perry has also produced an astonishing number of successful, multi-season TV shows in a relatively short period of time. Between 2006 and 2022, he created, wrote, directed, and produced 13 shows. Whereas many programs discussed in this book are limited series or run for just a few seasons, with even the most successful usually ending after four seasons, several of Perry's shows have had longer runs: *If Loving You Is Wrong* (5 seasons), *Meet the Browns* (5 seasons), *The Oval* (5 seasons and counting), *For Better or Worse* (6 seasons), *Sistas* (7 seasons and counting), *The Haves and the Have Nots* (8 seasons), and *Tyler Perry's House of Payne* (10 seasons). Seasons for Perry programs are also composed of more episodes than many others. Over the course of their four-season runs, for instance, the Georgia-made shows *Ozark* and *Devious Maids* included 44 and 49 episodes respectively, but the four seasons (and counting) of the show *Ruthless* (a spinoff of *The Oval*) includes 75 episodes. In total, Perry has made 1,536 episodes of TV shows, averaging 256 a year.

Perry maximizes profits by producing mass quantities of films and TV shows relatively quickly and cheaply. He often brags to interviewers that he can film three TV episodes in the time it takes Hollywood production companies to produce one. His industrial-sized production of product was satirized on an episode of *Atlanta* where a film studio head modeled on Perry cranks out reams and reams of script on a word processing piano that plays dissonant musical notes as he writes by pressing the keys (S4E5). In interviews, Perry is blunt about his focus on branding and profits, and a significant portion of his immense wealth comes from public coffers. In addition to the subsidy he received to purchase land for his studio, Perry benefits from state film subsidies. Since subsidies aren't capped in Georgia, the more shows he makes, the more wealth he transfers from social spending to himself. Tyler Perry Studios employ between 400 and 600 workers (many of them low-wage), which seems small relative to the breathless reports of Perry's billion-dollar fortune. Celebratory profiles deflect from the wealth disparities endemic to his industry by instead idealizing Perry as a financial giant. Media reports, combined with his own self-aggrandizement, glamorize the massive scale of his accomplishments in ways that connect him to the long history of US American idealizations of capitalist masculinity.[15]

Flight. Perry's investments in flying private jets suggest similar idealizations of class hierarchies and immense wealth. Jets are valuable materially—Perry's 14-seat Gulfstream jet reportedly cost $125 million, making it the third most expensive such vehicle in the world, and he also owns a larger 70-seat jet—but also symbolically, as signs of great privilege.[16] When asked by Trevor Noah to name "one thing I want to stay rich for," Perry said "just not to have to fly on a commercial plane is really cool. I know it's not a small thing, but it's really cool."[17] In addition to owning private jets, Perry is also a licensed pilot for a small passenger plane, which is prominently featured on his Instagram account. Just as he takes advantage of film and TV tax incentives, Perry enjoys a $1.8-million tax break from Cobb County, Georgia, in exchange for parking his jets at McCollum Field. The County justifies the tax break by citing the 10 jobs it generates at the airfield, but critics argue that's a poor return on such a large investment. As the *Atlanta Journal-Constitution* pointed out, "the amount the schools are giving up over the next 10 years . . . could pay the salaries of 28 teachers this school year."[18]

Perry further owns and flies hundreds of remote-control planes. Unlike the smaller RC planes widely available to consumers, Perry collects costly, custom-made planes that range in size from a compact car to a VW bus. His mansion includes an airstrip and hanger for his RC planes, as well as a statue of the Wright Brothers. "This is an expensive hobby," Perry explains in a flight demonstration video. As in accounts of his vast fortune, Perry seems to associate wealth with masculine size. In the flight demonstration video, for example, he tells his RC plane engineer Ramy to fly Perry's large replica of a Virgen Atlantic jet, but Ramy demurs, saying he's "good with my little cute sister," a smaller version of the plane, because "I don't think I have the balls to fly this." Perry responds "I got the Cessna over there and it was small, and I was like you know I'm having trouble seeing it. I want to build a big, big one. . . . So your wife says, she's behind the camera, she says yea, 'it's cute though' (mimicking a femme voice). (But) I don't want a cute plane; I want a *big* plane!" (in a booming, masculine voice).[19]

As Pierre Bourdieu argues in *Distinction*, like "all the metaphors of skimming and high flying suggest," wealthy people associate owning and flying a plane "with elevated society and high-mindedness, 'a certain sense of altitude combining with that of the spirit,' as Proust says apropos of Stendhal." Flying represents fantasies of upward mobility or what Bourdieu calls "social flying, a desperate effort to defy the gravity of the social field."[20] Interviews with Perry narrate his rags-to-riches story in terms of flight, charting his trajectory from living in his car or in a housing project by the airport to mansions and a private jet.[21] Flight further romanticizes extreme wealth by linking it to spiritual transcendence. As Perry told a reporter for *AARP* magazine, "humility stays very present because I remember [living] in that car. You know, I don't care if I'm on my plane at 40,000 feet in the air. I still remember those moments. And also, I understand that if something goes wrong at that 40,000 feet, ain't nobody goanna help me but God." Here he remarks on his poor beginnings to measure how wealthy he has become. At the same time, the experience of owning a private jet has seemingly afforded him a higher existential perspective on the world, at a dramatic distance from on-the-ground material necessities.

Ownership. Many of Perry's TV shows have screened on his friend and mentor's cable network, the Oprah Winfrey Network or "OWN," which is appropriate since "ownership" is a key term in his world. "OWN" suggests a double meaning—a station owned by a Black billionaire where regular Black people can watch stories about their own cultures and histories. Perry effectively subsumes the second meaning under the first, suggesting that his ownership of creative and material properties represents a collective win for Black people. He often tells the story of how his wealth was built by owning his own films and TV shows rather than selling them to others. As he proudly declares, "I own my brand."[22]

> Ownership is the key to make sure that longevity stays. I own everything. I would not sell a script. I would not sell a film. I would not sell a TV show. Nothing. I own it all. And to own a studio and to have *Black Panther* be shot there, which part of it was, and other movies be shot there. It's really, really phenomenal because ownership is the key to generational wealth, generational changes, and I think that's what we need to learn as people of color.[23]

Here Perry presents his enterprises as though they produce generational transformation for Black people as such. The quotation further suggests that owning a studio is especially important to Perry, and the fact that his studio was established on land that was once home to a Confederate military base partly built by slaves represents dramatic progress for himself, but also for other Black people. As he told the *AARP* magazine reporter,

> The land itself was once a Confederate Army base, which meant there were people here fighting to keep my ancestors enslaved. From the moment I walked onto the property, I was haunted by it. Sometimes when I'm walking here at night, I get a chill from all the things that have happened here. So, as we built each of the 12 soundstages, we buried Bibles underneath them, as a way of refocusing the spirit of the place. I wanted this to be a place where everyone was welcome.[24]

But the studio wasn't enough to satisfy his desire for ownership. "I got really depressed after my studio opened because I realized it was a major goal that I had obtained. I said, 'OK, now what?' I'd like to own a network. . . . I'd like to continue to grow this brand."[25] At the same time as he idealizes the ownership of private property and his brand, Perry also uses "own" in the collective sense, as when he tells an interviewer that Ray Charles, Billie Holiday, and Duke Ellington "were huge stars in their own community, you know, and that's pretty much my same story. I was able to build and have this amazing career among my own people."[26] Alternating between these two forms of "own," Perry implies that his wealth benefits all Black people.

Indeed, Perry argues that his property is an inspiration to poor Black people in Georgia. In his acceptance speech for a BET award, for example, he said, "When I built my studio, I built it in a neighborhood that's one of the poorest Black neighborhoods in Atlanta so that young Black kids could see that a Black man did that, and they can do it too."[27] He similarly claimed that one of his mansions was an empowering inspiration. As Kathleen Cross reported, "Perry says he believes his home should make a statement to those who doubt the power of faith and forgiveness. 'I want people to know what God can do when you believe.'"[28] Here Perry sounds like his friend and colleague the prosperity gospel televangelist Joel Osteen, intimating that his wealth is a sign of faith and divine favor.

Perry explains to the *AARP* reporter that his drive to own property originates in his working-class childhood: "My father was a subcontractor. He came home one day and was happy, because he had made $800 building a house. But he told me that the white man who owned the house later sold it for $80,000. That didn't make sense to me. I wanted to be the owner of the house." With this anecdote Perry recognizes the unfairness of capitalist exploitation and the violence of private property before drawing the lesson that it's better to be an owner than a worker. But ownership for some can mean exploitation and displacement for others. As Aymar Jean Christian and Khadija Costly White argue, Perry's claims to ownership are inflated and represent means of appropriating labor and hence profits. When four writers for his popular TV shows *The House of Payne* and *Meet the Browns* attempted to unionize and negotiate a contract with residuals and health benefits, Perry fired them. One of the writers, Christopher Moore, told entertainment journalist Nikki Finke, "It's very disheartening considering that this is a studio run by African Americans. What Tyler Perry is essentially saying to us is that 'you're black and there's not a lot of opportunities for you so you'll take what I give you'—whether it's fair or not."[29]

As a result, the four writers and other members of the Writers Guild of America (WGA) picketed the grand opening of Tyler Perry Studios. Celebrities including Winfrey, Will Smith, Sidney Poitier, Patti LaBelle, and others crossed a picket line to attend the event. Perry ultimately renegotiated a contract with the remaining writers, but none of them returned to the show and, as Christian and Khadija note, Perry has increasingly claimed sole authorship of his TV programs. Given the sheer number of works he claims to write, direct, and produce, they argue that it's hard to believe Perry's claims to individual authorship, while several accusations and lawsuits allege that he has stolen material from others. His ownership of both intellectual and material property, Christian and Kahdija conclude, enabled Perry to "bypass or accelerate what delayed and complicated production in the post-network era: the hiring of above-the-line creative talent, including writers, producers, and directors."[30] Similarly, Murali Balaji argues that he has pursued a profit strategy of employing nonunion labor and cutting the overall number of workers at his studio to maximize profits in ways that "help corporations minimize the costs of investing in the development of cultural products for African Americans." Perry's profits, Balaji concludes, uphold a "miserable hierarchy" comprised of "corporate winners at the top," and workers "whose ambitions and talents are exploited by corporate producers of culture" on the bottom.[31]

Finally, as George Lipsitz reminds us, public subsidies to produce Black entertainment commodities can steal educational opportunities from Black youth. Of public subsidies for the St. Louis Rams, Lipsitz writes that "cruelly enough, the success of Black athletes in St. Louis on the football field every Sunday helps build public identification with a project that systematically deprives Black children of needed educational resources."[32] Lipsitz's claim about the success of Black athletes also applies to stories about Perry's success, which, like football, is subsidized by state funds diverted from education.

Owners generally support the police, who are expected to protect their property, and Perry is no exception. In a local context where Black people are disproportionately subject to police violence and incarceration, and where community groups have protested the building of a new police training center colloquially called "Cop City" (see chapter 3), Perry has sided with the Atlanta Police Department (APD) and the mayor, who backs the center. In 2020, a month after the police murder of George Floyd, an APD police officer shot in the back and killed a Black man named Rayshard Brooks. Recalling the fictional efforts of Bishop Greenleaf, Perry helped local police with public relations, donating $50,000 in grocery store gift cards for officers to give away in the neighborhood where Brooks was shot. The APD Twitter account posted photos of police giving the cards to Black people and thanked Perry for helping them "spread good will."[33] The killing of Floyd further prompted Perry to defend the police in a speech at the 2021 Academy Awards that was viewed by over ten million people. Accepting the Jean Hersholt Humanitarian Award, which Perry told interviewer Joe Scarborough he believed the Academy awarded him in response to Floyd's murder, he used the massive platform to decry anti-police prejudice, as if the police were an aggrieved minority: "I refuse to hate someone because they are Mexican or because they are Black or white or LGBTQ. I refuse to hate someone because they are a police officer. I refuse to hate someone because they are Asian . . . I want to take this Jean Hersholt Humanitarian Award and dedicate it to anyone who wants to stand in the middle, no matter what's around the walls. Stand in the middle, because that's where healing happens."³⁴ Here Perry incorporates the police into a kind of identity politics of "the middle," where hierarchies and inequalities are dissolved into formal equivalencies, as on *Greenleaf*. Finally, in 2022, Tyler Perry Studios hosted the APD for an "Appreciation" picnic with food and live music. The event was attended by the Atlanta Mayor and Chief of Police, as well as local news media, and from the stage Perry declared to the audience of police officers, "We need you, the city needs you."³⁵ In all of these ways, Perry is a powerful agent of the kind of pro-police PR that defines the police and prison televisual complex.

Themes of *mass*, *flight*, and *ownership* are visible in one of Perry's most popular shows, *The Oval*, about the despicable Hunter and Victoria Franklin, the white US President and Black First Lady. While often referred to as "dysfunctional" by critics, the word hardly does justice to the first family's depravity. The Franklins constantly berate, demean, and physically attack their teenage children, Gayle and Jason. In one episode Victoria shoves her daughter out of a moving limousine and in another the First Lady tasers Gayle with a Secret Service agent's weapon. In the premier episode of season 2, Gayle is banished to her evil grandparent's country house for unnamed "punishments," never to be seen again. The same episode reveals that Jason has raped and murdered a White House maid and decapitated a Black woman with whom his father was having an affair. Meanwhile, Victoria arranges for a Secret Service agent to murder the President's Press Secretary and lover. Later, the First Lady attempts to murder both her own son, to prevent him from talking to the press, and then her husband, to replace him with her lover, Vice President Eli. For his part, President Franklin holds the responsibilities of the job in contempt, spending his days and nights drinking scotch, snorting cocaine, and visiting sex workers, enlisting a Secret Service agent to act as his drug dealer and pimp. In one episode he tasks an agent with kidnapping a pharmacist and bringing her to the White House so Franklin can coerce her into becoming his mistress (S3E8). He also threatens to have her boyfriend murdered, and even threatens the children of the Vice President. Finally, when the President learns that the First Lady tried to poison him with fentanyl, he orders a secret service agent to torture her.

The Oval is outsized by many measures. Seasons are 22 episodes long, more than double the size of many contemporary series, with an expansive cast of characters. In addition to the first family, the show follows the complicated lives of White House workers and their kin. One group is represented by butler Richard Hallsen, his wife Nancy, his son Barry, Barry's ex-girlfriend Sharon (the pharmacist propositioned by the president), and her current boyfriend, pharmacy manager cum drug dealer Kareem. Second is Donald Winthrop, White House chief of staff, his wife Lilly Winthrop, the First Lady's stylist, and the couple's respective lovers Secret Service agents Kyle and Bobby. Priscilla Owens, White House residence supervisor, and her Secret Service agent husband Sam form the third set of characters; when Sam has an affair with the First Lady, Priscilla romantically pursues his fellow agent, Bobby. The fourth character grouping includes Vice President Eli and the Second Lady, Simone. As these four groupings suggest, the large cast of characters is narratively organized into a profusion of love triangles connecting central and more minor characters. When agent Kyle flirts with Sharon's pharmacy co-worker Dale, Chief of Staff Winthrop threatens to kill him; a policeman falls in love with Lilly and tries to murder her husband, Chief of Staff Winthrop; the chief of staff's assistant Dale is the jealous ex-boyfriend of Elle, the press secretary sleeping with the president, whom Dale tries to kill after her death. The kaleidoscopic love triangles produce seemingly inexhaustible narrative possibilities. Perry's prodigious narrative productivity is overflowing, generating, for example, a successful spinoff titled *Ruthless* about a murderous sex cult and featuring Ruth, a minor character from *The Oval*.

Perry is fond of saying that the broadly entertaining aspects of his work, especially its humor, help draw audiences into serious narratives about abuse, infidelity, forgiveness, and love. Recalling his own childhood traumas, he often dramatizes abusive relationships between parents and their children. But *The Oval* is all violence and abuse without any grace notes of redemption. The program's engagement with the symbolism of flight suggests a similar interpretation. Tyler Perry Studios employs a large commercial airliner set, constructed out of two decommissioned planes. On *The Oval*, the set was used for a shocking scene where first son Jason strangles to death his grandmother. In contrast with representations of flying in terms of transcendence and upward mobility, here the plane is a setting for murder. Finally, home ownership on *The Oval* makes murder possible. Perry is the proud owner of a permanent, 80% scale reproduction of the White House built for *The Oval*. It has three floors, working plumbing, and a portrait of President Barack Obama. As Perry told an *Architectural Digest* reporter, "this is the only White House that has ever been built at a studio on a lot."[36] The Obama portrait notwithstanding, *The Oval*'s White House is no space of hope and change. Instead, it is a setting for murder, rape, bloody fights, torture, and imprisonment. In my reading, the show thus represents the unconscious of Perry's self-promotional projection of massive success. *The Oval* is obsessed with mass violence, as if symbolically referencing its dependence on the violence of redistributing potentially life-sustaining social spending from poor and working people to the wealthy.

In *The Oval*'s world, the police, including the Secret Service, are agents of violence and death. As various characters discover, DC police are in the pocket of the President and the Secret Service. When Kareem tells the police that the President kidnapped Sharon, they inform an agent who then threatens to torture Kareem. Meanwhile, agent Kyle murders multiple people execution-style, including another agent. He also waterboards his lover's wife, Lilly. Considering Perry's support for local law enforcement, such representations seem like pandering to Black audiences with an understandable antipathy toward the police. At the same time, *The Oval*'s police can be read as sensational representations of the carceral worlds that TV production in Atlanta supports. As I argued in chapter 3, TV studios and the police in Georgia are allies. They share state funding models and occupy state-subsidized land in similarly destructive ways for nature and low-income communities. Hollywood supports Georgia prisons and jails by filming in them, and by presenting generally positive, heroic representations of the police. By contrast, *The Oval* represents the violent underside of the more commonly idealized relationship between TV and the police. As such, Perry is a major player in the police and prison televisual complex.

TV Shows about TV Shows: Mara Brock Akil's *Being Mary Jane* and *Love Is_*

While not as prolific as Perry, the writing/directing/producing team of Mara Brock Akil and her husband Salim are prominent creatives in Atlanta-made television. Brock Akil was a writer for four seasons on the UPN situation comedy *Moesha* (1996–2001). She also created and wrote two spinoffs, *Girlfriends* (UPN, CW, 2000–2008) and *The Game* (CW, BET, 2006–2015). In 2010, production for *The Game* shifted from Los Angeles to Atlanta for six seasons. Salim Akil was a writer and director for *Girlfriends* and *The Game*. He also created, wrote, and directed the Atlanta-set and -shot superhero show *Black Lightning*. (See chapter 4.) She wrote, he directed, and together they produced the show *Love Is_*, filmed in Atlanta but based on their own romance while working in the 1990s Los Angeles TV industry. Finally, Mara Brock Akil created and helped write *Being Mary Jane*, while Sam Akil directed 17 out of 45 episodes. The show was very popular, averaging 2.6 million viewers per episode.[37]

With *Being Mary Jane* and other shows, Mara Brock Akil has positioned herself as a spokesperson for DEI TV. Eschewing colorblind casting, which results in characters written to be generic and hence effectively white by default, Brock Akil argues that her shows are "Black on purpose."[38] With speaker's fees of $30,000 to $50,000, she addresses corporate and university audiences on topics such as "Breaking Barriers and Empowering Others," a talk advertised in this way on a speaker's bureau website:

> MARA BROCK AKIL revolutionized television by creating a black space for entertainment. In these illuminating and powerful remarks, she offers valuable insight into creating diverse and inclusive spaces where every community-member feels empowered to be their authentic selves. Shedding light on how she's broken barriers in Hollywood and leveraged her platform to empower women globally, Akil's remarks balance bold inspiration and actionable lessons that leave audiences ready to speak up and make change.[39]

As we will see, with some slight changes this could be the description of a talk by Brock Akil's most famous character, Mary Jane Paul, who also makes TV shows.

Being Mary Jane is about the romantic, familial, and professional challenges faced by an Atlanta TV news anchor Mary Jane Paul (Gabrielle Union). Her father, Paul Patterson Sr. (Richard Roundtree) is the retired CEO of Vantage Airlines, and her mother, Helen Patterson (Margaret Avery) is a prominent member of Atlanta charity boards. Mary Jane and her brothers Patrick (Richard Brooks) and Paul Jr. (B. J. Britt) all graduated from Morehouse. Patrick, his daughter Niecy (Raven Goodwin), and her two small children live in the Pattersons' three-story colonial style brick home. Mary Jane, meanwhile, lives in a gated midcentury ranch home with a pool and floor-to-ceiling glass windows. In the first season, actual Atlanta homes were used as locations but in subsequent seasons home interiors were filmed on soundstages, including in a scale recreation of the house used as Mary Jane's home in season one. The glass-walled house and set design were selected to suggest that, as a local TV personality, Mary Jane lives in a fishbowl. Many scenes were shot from outside the home looking in, visually evoking the experience of a Black woman under a TV gaze that has historically been dominated by white men. As previously noted, *Being Mary Jane* presents critical reflections on the combined racism and sexism Black women encounter in the media and other walks of life, as well as white nationalism and police violence, but the show does so through a decidedly upper-class respectability politics lens.[40]

As Natasha R. Howard argues in her essay "Real, Respectable, or Both: Respectability on *Being Mary Jane* through the Words of Mara Brock Akil," the character's "choices reflect her adherence to respectability expectations as a successful Black woman journalist and for her family."[41] With her multiple sexual partners and extensive drinking, Howard argues, the character departs from gendered norms of respectability, but in other ways Mary Jane exemplifies upper-class respectability. The show, for example, emphasizes her cleanliness, repeatedly depicting Mary Jane cleaning her home, even though she also employs a Mexican housekeeper. The premise of the program is that while successful at her job, she longs for marriage and children. Mary Jane's respectability politics also include a commitment to racial uplift. "As a journalist with a desire to bring attention to issues that matter to her and are related to the Black community, community uplift is mainly represented while Mary Jane is at work," Howard writes. Mary Jane also, she argues, expresses ideals of community uplift in conversations with her siblings, whose self-improvement, education, and gainful employment she forcefully promotes.

Being Mary Jane uses conflicts in the Patterson family to investigate class differences, with the characters Patrick and Niecy representing "low class," nonrespectable lifestyles. Patrick is a recovering drug addict whose girlfriend Tracy, a working-class white woman and mother of his children, is also in recovery. At the start of the show, he is unemployed and asks his wealthy sister for a loan, much to her chagrin; Mary Jane also criticizes him and Niecy for living off her parents. Niecy has two children by two absent fathers. Her aunt constantly lectures her about not getting pregnant, about getting a job, about how she dresses, about how she spends her money, and even about her weight, since Niecy is larger than the thin ideal Mary Jane and her professional friends represent. Indeed, the show maps class differences onto differences of female size and weight. Low-class characters are often represented by actors larger than more classy characters, including Niecy, Tracy (who confesses to trading sex for drugs), and the show's most significant villain, Cece (Lorretta Devine), a masculine-presenting queer character who attempts to extort Mary Jane. As Howard concludes, "when it comes to respectability, the inner workings of how important image is to those that uphold respectability politics is one that Brock Akil enforces in Mary Jane's world. Physical image and reputation are everything" to the character.[42]

Being Mary Jane suggests that Niecy's bad choices about sex and her weight are connected to other poor choices. In one narrative arc, she is stopped by police while driving and is tasered, and as a result she's awarded a $150,000 settlement from the Atlanta Police Department (S4E4). The brief focus on police violence is quickly displaced, however, by scenes of Niecy spending the settlement money in ways that the show depicts as irresponsible. She initially plans to put it in a trust fund for her children but when one of their fathers pressures her for money, she buys him a red sports car. Niecy also buys herself expensive, garish fashions and pays for a VIP booth at the club for her friends, but she gets in a physical fight there with a woman who tries to crash her party. The fight is captured on cell phones and goes viral. In response, Niecy discusses her case on social media, which violates the terms of the police settlement, so she's forced to pay the money back. Here the show exploits the topical issue of police violence to vilify a poor Black single mother, one of the demographics most disadvantaged by state spending priorities that divert public resources to the wealthy.

For the most part, however, *Being Mary Jane*'s DEI representations deny both its classist narratives and the show's reinforcement of class differences via state subsidies for TV production—with one glaring exception. Under pressure from Mary Jane, her brother Patrick takes a job guarding local film and TV production trailers. Although it's one of the employment opportunities promoted by boosters of Georgia's filming incentives, Patrick's position is depicted as shameful. When he goes golfing with his CEO father, an acquaintance recognizes them and asks Patrick what he's doing for a living; when Patrick looks embarrassed, Paul Sr. quickly changes the subject (S2E3). As Patrick later complains to his parents, "This job, Lordy. They got me on call for 25 lousy hours a week. They never tell me when I'm working so I can't plan my schedule, and it's tough finding sitters at the last minute." At a dinner party, standing in her grand kitchen, Mary Jane incredulously tells her guests that their family "used to have a spirit about us, and we need that spirit back because we have a brother, who also went to Morehouse, who babysits movie trailers for a living." The camera then cuts to Patrick at night, in a security guard uniform, walking past rows of trailers, as Mary Jane continues her speech on the soundtrack: "And I can't enjoy any of this, I can't stop and smell the roses, knowing his ass is sitting out there" (S2E1). While the dialogue and editing distance Mary Jane's expensive home from the space of low-wage film production labor, her voice on the soundtrack bridges the cut and connects the two scenes, as if suggesting that Mary Jane's TV wealth depends on the exploitation of low-wage workers like her brother. Responding to a journalist's questions about the scene, Brock Akil explained that she identified with Mary Jane's predicament: "I find even personally, sometimes it's hard to be successful when your community or your family is struggling. That's what you're seeing in that scene. . . . That's what's on our minds, trying to feel okay with your own success. More specifically for Mary Jane, with her own brother."[43] Brock Akil suggests that her enjoyment of her own success would be compromised if a member of her family or community was employed in one of the low-wage jobs TV production brings to Georgia.

TV shows made in poor places, I argue, represent their own mode of production in displaced, symbolic forms. TV shows within TV shows, like the one on *Being Mary Jane*, invite interpretations about the process of producing TV under local conditions of racial capitalism. One of the clearest, most revealing examples of a symbolically significant show-within-a-show is Mara Brock Akil and Salim Akil's *Love Is_*, which is self-consciously based on the couple's work history and romance in late 1990s TV. The character based on Salim is an aspiring screenwriter named Yasir (William Catlett), while the character based on Mara is an established Black sitcom writer named Yuri (Michelle Weaver). Her success has enabled Yuri to buy a home, a Spanish-style bungalow prominently featured on the show. In the first episode she tells two other writers that as a homeowner, she wants a man with similar professional accomplishments and financial standing. In the following writers' meeting, the head writer Norman (Kadeem Hardison) warns her that unless she contributes more jokes, she might lose her house. As in Perry's world, the idealization of Black home ownership is part of the self-justifying appeal of working in TV, despite or perhaps even because as a force of gentrification, TV production threatens to displace others from their homes.

The writers' room is one of the show's most important sets, a conference table topped with pens and containers of Twizzlers, with a whiteboard and a cork board covered in Post-it notes. The camera often circles around the table to showcase the all-Black writers' room, six men and two women, Yuri and Angela. Different episodes find the women exchanging knowing glances as the men make sexist jokes, and their efforts to write better stories and dialogue for Black women characters are often resisted. But *Love Is_* presents a generally positive, nostalgic representation of the writers' room, complete with popular hip hop music from the 1990s on the soundtrack. Showrunner Norman is a gruff but supportive mentor for the women writers, collaborating on a show for Whitney Houston with Angela and helping Yuri produce her first script. Norman is played by Kadeem Hardison, whose star image enhances the character's warm aura. Early in his career, Hardison co-starred with Jasmin Guy and others in the popular HBCU comedy *A Different World*; like the soundtrack, his casting in *Love Is_* contributes to its '90s nostalgia. More recently, Hardison has played gruff mentors to young women not only in *Love Is_* but also in *Teenage Bounty Hunters*, where he plays a bounty hunter who shows two white girls the ropes of the job (chapter 3). The idealization of the showrunner and TV writers' room on

*Love Is_* distances it from the program's conditions of production, with their basis in the appropriation of public resources diverted from spending on housing and education. The representation of a Black writers' room thus doubles for the actual writers' room that produced the show.

In contrast with the supportive mentoring relationships between showrunners and staff writers on *Love Is_*, Brock Akil has been accused of plagiarizing from other Black women writers. In 2006, Staci Robinson sued Brock Akil, claiming that after Robinson sent Akil a copy of her novel as part of her application for the job of writer's assistant, the showrunner used it as the basis for TV show *The Game*. In 2007 Brock Akil settled the lawsuit for an undisclosed sum of money.[44] Similarly, in 2018, writer and actor Amber Dixon Brenner sued Salim Akil for domestic abuse during their affair and sued the couple for stealing elements of her screenplay for *Love Is_*. Salim Akil has denied the charges, but in response to the lawsuit, OWN cancelled *Love Is_*. The show, however, displaces any possible references to plagiarism and abuse with the representation of Norman, the caring showrunner. Just as it displaces plagiarism with representations of mentorship, the program similarly displaces the appropriation of public resources. Ultimately, *Love Is_* suggests, contra the forms of racial capitalism on which it depends, that TV production is a fair, equitable, and attractive form of work.

Respectability TV

Alexander-Floyd argues that one contemporary legacy of the Moynihan Report in politics and popular culture is "a middle-class 'ideology of respectability' centered on restoring two-parent, patriarchal homes in black communities (and) promoted as the solution to black social and political ills."[45] Representations of Black respectability in film and TV "juxtapose two black Americas—that is, the underclass, ridden with moral failings that drive their poverty and other social ills, and the middle and upper class, whose members, the most noted being Barack and Michelle Obama, are held out as moral exemplars for their less successful counterparts." While neoliberalism is a later historical development, Alexander-Floyd argues that respectability politics promote neoliberalism by celebrating individual responsibility and patriarchal families as the horizon of political aspirations and demands. Melodramatic narratives help narrow political horizons by focusing on conflicts involving individuals, romantic relationships, and families. From that perspective, the answer to inequality lies not in activism or state actions but in respectable conduct. Black melodramas are "undemocratic," Alexander-Floyd suggests, because they encourage viewers to see the causes of and solutions to social problems in terms of individuals and families, "as opposed to macro-level structures or a broader set of political and social forces." She concludes that "it is no accident that we see a plethora of narratives focused on self-rehabilitation and channeling social concerns into the realm of the family. The neoliberal focus on self-making and refashioning is an important accompaniment to the retreat of the state that facilitates the expansion of global capital unfettered by government oversight or responsibilities to taxpayers."[46]

What Alexander-Floyd argues about TV programs such as *The Cosby Show* and the films of Tyler Perry is also relevant to the Atlanta-made shows analyzed in this chapter, with an additional layer of significance. Representations of Black respectability justify the political economy of local TV production based in corporate subsidies and the upward redistribution of wealth from poor Black people to more privileged Black creatives. State programs that divert public resources from social welfare to Black media makers exemplify the retreat of the state, freeing racial capitalism from government oversight or responsibility to local people where TV shows are made. State subsidies for TV production in Georgia thus suggest comparison to the history of structural adjustment programs in which international financial institutions pressed elites in the global south to cut social spending and regulations to make their countries more attractive to investors. Georgia state subsidies reterritorialize structural adjustments locally, making poor Black people of the state part of the global south in the north. Taken together, the TV shows featured in this chapter obscure, deny, and justify such forms of Black racial capitalism in place.

To return to *Greenleaf*, one of the show's narrative arcs represents the effects of TV racial capitalism in displaced form, giving viewers a glimpse into the real consequences of state subsidized TV for working-class Black people. Produced for the Oprah Winfrey Network, Winfrey herself took a recurring role on the show as Mavis McCready, Lady Mae's sister. Mavis is an outcast in her church family who owns a blues

Figure 6.4: Mavis McCready behind the bar in *Greenleaf*.

club. When Grace visits her aunt in her club, she recalls that her mother, Mae, the bishop's wife, had told Grace that her "evil sister" Mavis "played the devil's music." Resembling *Being Mary Jane*, the show maps class differences onto family differences in the juxtaposition of Mavis and the Greenleafs. Winfrey's character is depicted as a brave truth-teller who confronts traumatic family secrets, but from an unrespectable and "low class" position. Mavis's most common setting is behind or at the bar in her blues club, but in season 1 the city seizes the club through eminent domain to turn it into a museum dedicated to an abolitionist newspaper from the 1850s. (See Figure 6.4.)

The seizure is part of the mayor's effort to generate good will among Black voters in the wake of the police shooting of the unarmed Black teen I discussed at the top of the chapter. In a turn seemingly inspired by state subsidies for TV production in Georgia, the appropriation of a Black woman's property funds a DEI museum that smooths over police violence. The character of Mavis in this scenario represents the global south in the north reproduced by TV racial capitalism. *Greenleaf*, in other words, gestures toward the police-aligned racial capitalism in place that is so often disavowed on TV programs.

7

Gentrification TV

Like the George Romero film of the same name, the Atlanta-shot anthology horror show titled *Creepshow* is inspired by EC Comics from the 1950s. In an episode called "Pesticide" (S2E2), a mysterious real estate developer named Murdoch (Keith David) hires an exterminator named King (Josh McDermitt) to address his "infestation problem" by poisoning homeless people living in an old warehouse that Murdoch plans to turn into condos. After completing the job, King is hounded by giant insects (real or imaginary), and when he falls asleep in a client's home he wakes to find that he has shrunk to the size of an insect, and the client then squashes him with a rolled-up magazine. "Pesticide" thus mimics EC's "snap ending," or what comic scholar Qiana Whitted analyzes as the "surprise plot twist" that brought many EC stories to a close. Snap endings represent poetic justice, where the punishment parallels the crime. Or as EC publisher William Gaines explains, "If somebody did something really bad, he usually 'got it.' And of course the EC way was he got it the same way he gave it."[1] So in "Pesticide," King, who treats the homeless like insects, is himself squashed like a bug.

The episode "made a big political statement," according to the actor who played Murdoch. As David explained, "The reason why we have so many homeless people is because they're almost seen as outcasts. People like Mr. Murdoch who have the wealth and means to help other people do better don't [because] they just want to keep it to themselves. So, it's easier for them to get rid of them and treat them like vermin than to treat them like human beings."[2] David's interpretation of the episode draws our attention to its divergence from EC conventions. The "snap ending" formula defines King's character arc, but Murdoch escapes the circle of poetic justice. In the final scene of the episode, Murdoch reappears at the client's door dressed in an exterminator's uniform and asks with a grin, "Bug problem?" As David suggests, the developer is the source of the horror but only his employee King is subjected to horrifying poetic justice.

Here Murdoch is a kind of displaced stand-in for film and TV studio developers. The character aims to convert an old warehouse into upscale housing. As I argued in chapter 2, TV warehouses can be read as indirect representations of studio soundstages, and contemporary studios often also include housing developments. By letting the real estate developer off the hook, *Creepshow* is emblematic of Hollywood's depiction of gentrification. Like the character of Murdoch, TV producers absolve themselves of blame, effectively denying their own gentrifying effects in the poor places where they film.

In this chapter, I investigate how studios make TV shows critical of gentrification that, in their mode of production, abet gentrification. Film and TV studios are effectively forces of gentrification, building wealthy residential developments with public subsidies that could instead pay for low-income housing and other social services. Studio redevelopment raises surrounding property values, displaces residents, and encourages expanded policing.[3] Public/private partnerships have diverted funds from social spending in urban redevelopments anchored by film studios and adjacent mixed-use residential developments. I thus analyze shows about housing developments as displaced interpretations of gentrifying TV production.

Read against the grain, programs about gentrification draw critical attention to TV production as a mode of gentrification. Viewed in critical juxtaposition, they also reveal that the contemporary wave of TV gentrification is another name for settler colonialism. With its satire of gentrifying "white saviors" in settler contexts, *The Curse* bears comparison to *The Walking Dead*, with its white policeman savior, and to *Vida* and *Gentefied*, which include gentrifying brown saviors. The last two shows, produced by brown creatives, thus represent examples of Latinx TV racial capitalism in place. Finally, the juxtaposition of New Mexico and other locations invites investigation into the occluded history and ongoing reality of film and TV production as a settler colonial project, not only in terms of representation but also via the occupation of Indigenous land, including in Los Angeles.

Focusing on a particular facet of TV racial capitalism in place, in the next section I provide an overview of gentrifying TV architectures and their encouragement of expanded policing among poor people of color in Georgia, New Mexico, and East Los Angeles. To compete with states and countries that offer significant subsidies, the state of California and City of Los Angeles provide extensive resources to film and TV producers that drives gentrification. In the remainder of the chapter, I present a comparative study of shows about gentrification in Georgia (*The Walking Dead*), New Mexico (*The Curse*) and the Boyle Heights neighborhood of East Los Angeles (*Falling for Angeles*, *Vida*, and *Gentefied*). A recurring thread in this book connects diversity in TV content, writers, and actors to the conditions of TV production, or what I theorize as DEI TV. *The Curse*, for example, critically reflects on white gentrification as a form of settler colonialism. *Vida* and *Gentefied* were produced by Latinx creatives, writers, and cast and they present complex representations of Latinx race, sex, and class differences, including conflicts and contradictions within Latinx communities. And yet via the system of state subsidies for TV production, all the shows in this chapter contribute to the kinds of gentrification they depict, a situation obscured by their diversity.

TV Gentrification

This section builds on Vicki Mayer's *Almost Hollywood, Nearly New Orleans: The Lure of the Local Film Economy*, where she argues that, in tandem with tourism, state-subsidized film production has helped gentrify New Orleans.[4] As she writes, "proselytizers for film economies tout the use of public funds to return 'undervalued properties' to the real estate market, where they can find new economic life for private investors. By offloading their risk onto the public, the investors can look to profit by building production infrastructure and 'loft-style living' accommodations for film personnel." As a result, many corner stores and dive bars in New Orleans have been replaced by upscale cafes and cocktail bars. Mayer concludes that the "vast economic disparity between those who get public subsidy through the film economy and the vast majority of working-class people who do not is nowhere more evident than in housing, where gentrification has transformed entire neighborhoods."[5]

Film and TV production in New Mexico and Georgia contributes to gentrification in similar but even more extensive ways, with film studios supporting affluent communities. Several contemporary studios, for example, incorporate mixed-use real estate developments. In New Mexico,

Figure 7.1: Netflix's Mesa del Sol housing development, with soundstage 3 visible in the distance.

Albuquerque Studios was already connected to a development called Mesa del Sol when Netflix purchased it in 2018. (See Figure 7.1.)

The streaming giant is in the process of a massive expansion that will add over 900 acres. The site includes, or will include, smart suburban homes, parks, trails, restaurants, an international school, a soccer stadium, and the Del Sol Church. The studio has been home to *Breaking Bad*, *Better Call Saul*, and a host of other shows. More recently, the Netflix action series *Obliterated* (2023) used the streets of Mesa del Sol to film a car chase (S1E6). The studio housing development is part of a larger patchwork of segregated developments dividing Albuquerque along class lines, suburban sprawl verses urban and rural poverty. Such spatial divisions are visible, scholars argue, in *Breaking Bad*'s contrasting suburban and urban locations.[6]

Trilith Studios (formerly Pinewood) in Fayetteville, Georgia, half an hour south of Atlanta, is a 935-acre development that includes filmmaking facilities and a 235-acre "European" town with homes, lofts, restaurants, shops, parks, trails, a lake, schools, a fitness center, a spa, and a branch of the Passion City Christian Church. (See Figure 7.2.)

"This is the future of filmmaking and content producing," the studio's website announces, "where the world's best talent live and work side by

Figure 7.2: Trilith Studios and housing development.

side."[7] According to *Variety*, "The single-family homes are priced around $700,000 and up—more than twice what homes go for in surrounding neighborhoods, but reasonable by L.A. or N.Y. standards. . . . That means that while actors, catering managers and crew members could potentially afford to live there, it's unlikely that the maintenance crew would be able to."[8] Despite lip service to diversity, it is limited by the high cost of housing but also by reports of discrimination. Trilith is the dream of Dan Cathy, former Chick-fil-A CEO. The far-right billionaire is infamous for bankrolling electoral efforts to ban same-sex marriage, making the Town at Trilith uninviting for queer people. As a reporter for *BuzzFeed News* wrote, most of the studio and Town employees are white "and cut from the same conservative, religious cloth." In that context, the small number of Black residents report experiencing racism from town leadership, neighbors, and the police. In a 2022 lawsuit, Black residents charge that town leaders treated them differently than white neighbors, making them wait longer for repairs and forbidding structural modifications allowed for white homes. White neighbors "policed" supposedly common space, publicly declaring that people of color were outsiders who didn't belong in pools and on playgrounds and calling local police. When asked by a reporter about the racial profiling of a young Black resident, police affirmed their duty to prevent the unauthorized use of the development's "amenities."[9] Although the plaintiffs do not feel at home there, Trilith is the production home for many Marvel movies and

TV shows with diverse casts, including *Black Panther, Black Panther: Wakanda Forever* and *Echo* (about a deaf and disabled Choctaw woman hero, and with a large Indigenous cast and director). In the words of the discrimination lawsuit, "The Defendants failure to address issues of race discrimination and inequity demonstrate hypocrisy within the community where Trilith promotes values of diversity and inclusion for purposes of profiting while simultaneously condoning discrimination against the black residents living within its community."[10] The lawsuit suggests that Trilith Studios and housing development is a concrete example of my claim that DEI TV representations conflict with and help hide exclusions and inequalities at the site of TV production.

Elsewhere in Georgia, another mixed-use community is in the works at Assembly Atlanta, the old auto factory converting to a studio. Once completed, it will include "state-of-the-art movie television and movie production studios" and a mixed-use development with apartments and townhouses, a 5-acre park, entertainment venues, and a boutique hotel.[11] And in 2023, Tyler Perry purchased 37 more acres of Fort McPherson land adjacent to his Atlanta studio to build an entertainment and residential district with a theater, apartments and condominiums, retail shops, restaurants, and green spaces.[12] But the oldest and most successful TV studio and mixed-use development is in Senoia, Georgia, about an hour southwest of Atlanta. Most interior scenes for *The Walking Dead* were filmed at Raleigh Studios there, but many exteriors were shot in the town itself. In 2017, AMC purchased the studios for $8.2 million. Raleigh's president before the sale, venture capitalist Scott Tigchelaar, is also the current president of Senoia Enterprises Inc., which redeveloped historic downtown Senoia. Among the recent additions is Nic & Norman's, a restaurant named for *The Walking Dead* director Greg Nicotero and star Norman Reedus. (Nicotero is the executive producer of *Creepshow*, and he directed the anti-gentrification episode "Pesticide.") Senoia has thus become a tourist destination and gentrifying suburb. AMC built several houses in Senoia that served as locations for the community of Alexandria in *The Walking Dead*. (See Figure 7.3.)

Once the show wrapped, the homes were listed for sale for $800,000 to $1 million, and more homes were built in the vacant lots where *Walking Dead* sets once stood.[13] Tigchelaar has proposed a trolley system to connect Senoia to Trilith, and to Bouckaert Farm, an upscale equestrian

Figure 7.3: Alexandria, *The Walking Dead* housing development.

center, wedding venue, and film and TV production location set amid forests, lakes, and the Chattahoochee River.[14]

Research on the greater Atlanta region suggests that studios are located in lower-priced areas where their presence increases home values.[15] Since 2010, development-driven increases in property values in Atlanta have led to the displacement of low-income residents and the increased whitening of neighborhoods with significant Black populations, a trend to which the studios contribute.[16] Rising property values and displacement undermine studio DEI initiatives, since gentrification excludes Black people from the area before they can benefit from such programs. As Emily Wood of the organization Defend ATL told a reporter, "the majority of these jobs will go to people outside the community with specific skill sets. That will increase property values and, without rent control, will push residents out."[17]

Tyler Perry Studios is a prime example of the dynamics analyzed by Wood. As noted in chapter 2, film and TV studios in Georgia raise property values and taxes, displacing existing residents, and excluding potential future residents. In recognition of the fact that his studio has displaced poor Black seniors, Perry has donated $2.75 million dollars over five years to help defray the costs of increased property taxes. The donation is notable but modest relative to profits (Perry is a billionaire who made $175 million dollars in 2023) and it only addresses the needs of a small number of existing homeowners and does nothing for the potential homeowners excluded from the area around Tyler Perry Studios due to rising property values. We are left to wonder how housing access would be different if subsidies for film and TV production were instead directly invested in low-income housing.

The security features of studio housing developments complement Hollywood investments in the criminal justice status quo described in previous chapters. Adjacent to Albuquerque's South Valley, a poor area around 80% Latinx, the Netflix housing development is advertised in terms of protection from crime. In the name of public safety, Mesa del Sol's developers have limited access points, incorporated "asset tracking technology," and fostered "impactful relationships" with the Albuquerque Police Department and the Bernalillo County Sheriff's Department to ensure regular area patrols. The development also employs a private security force.[18] Similarly, the Trilith and Senoia developments emphasize their distance from Atlanta and, implicitly, from urban crime. Trilith is "a perfect haven for those wanting to escape city life for a more peaceful existence... in a quiet haven separate from Atlanta's urban environment."[19] The website for the Senoia development calls it "a special place... frozen in time... Senoia is a destination for those who seek to decompress from a modern, fast-paced lifestyle, or escape it altogether. We welcome you to a passage back to simpler times, where neighbors are friends, shopkeepers believe in service, and family values are important to everyone."[20] As I noted in chapter 3, by renting police vehicles, *The Walking Dead* enabled the Senoia Police Department to purchase and support a K-9 unit, while the Department's web site seems to advertise and promote Senoia PD by featuring a large background photo of a part of town *The Walking Dead* famously used to play a settlement of the living. Which is to say that gentrifying housing developments are part of the police and prison televisual complex.

The Boyle Heights area of East Los Angeles has a long history of residential displacements, culminating in contemporary forms of state subsidized gentrification. Boyle Heights historian George J. Sánchez demonstrates that in the twentieth century, the neighborhood was hit with three waves of removal that anticipate the more recent displacements produced by gentrification. He charts the Southern California history of restrictive housing convents which helped make Boyle

Heights a racially mixed neighborhood filed with Mexicans, Japanese, and Jews excluded from other parts of Los Angeles. City officials projected ideologies of white supremacy and racial inferiority onto regional social space, paving the way for the displacement of Mexicans from the neighborhood during the era of California's mass deportations in the 1930s, the internment of Japanese residents during World War II, and the displacement of working-class people of color due to urban renewal and the building of freeways in the 1950s and 1960s. As Sánchez explains,

> A certain ideology developed among city leaders and urban planners that joined local politicians and bureaucrats on both the conservative and liberal sides of the political spectrum in the region, linking racial depravity and urban space. This ideology associated particular neighborhoods like Boyle Heights with slum conditions and urban decay, and it prompted local officials to consider residents of these neighborhoods as utterly (re)movable in order to make way for their plans to improve social conditions and urban progress.[21]

The racist ideology governing earlier forced removals continues to shape contemporary gentrification in Boyle Heights, where the forces of redevelopment implicitly draw on historic images of the neighborhood's people as barriers to progress.

For the past 20 years, the Los Angeles city government has invested more than $3 billion in Boyle Heights to attract capital. City investments in the neighborhood are part of an "aggressive arts-oriented development," including a new arts district. City redevelopment projects have attracted art galleries, cafes, restaurants, and condos. Although on a smaller scale than in Georgia and New Mexico, several film and commercial studios have opened in Boyle Heights. Monarch Studios, for example, advertises its "large industrial warehouse, a cyclorama infinity wall, and various standing sets for photo and video shoots" such as a back alley "Street Scene," perfect for inner-city crime narratives. Rappers Lil Wayne and Baby Tate have filmed videos there.[22] "A studio built for the creator generation," Pollution Studios produces commercials for clients such as Disney, GQ, and Perrier. "Guardians of Life," a PSA about threats to the Amazon rainforest starring Joaquin Phoenix and Rosario Dawson, was also shot there.²³ Together, these different forms of gentrification in Boyle Heights have displaced an estimated 2,500 local families.²⁴

In response, local activist groups such as Defend Boyle Heights, Boyle Heights Against Artwashing and Displacement (BHAAD), and the Ovarian Psycos have organized against gentrification.²⁵ While such groups have been the focus of media coverage, anti-gentrification efforts have also been organized by Latina immigrants living in public housing as part of the L.A. Tenants Union.²⁶ The term "artwashing" refers to the use of state-subsidized art developments to justify and distract from gentrification. As Kean O'Brian, Leonardo Vilchis, and Corina Maritescu write,

> Artwashing presents gentrification as beneficial to communities (variations on "improving the artistic life of a neighborhood") while ignoring the material impacts and effacing the actual needs of the neighborhood (Boyle Heights, for one, needed job-providing factories, grocery stores, and laundromats more than it needed galleries). The state displaces low-income folx, immigrant families, and other vulnerable communities under the cover of "building arts districts," or "river revitalization" efforts, together with public investment and tax subsidies, and the development of so-called affordable housing and luxury living.²⁷

Along with galleries, I argue that TV production in Boyle Heights is an influential form of artwashing that obscures its gentrifying effects.

Long a Hollywood location for filming police dramas about Mexican gangs, several TV programs have used Boyle Heights for crime scenes. On *Gang Related* (Fox, 2014), a show about a Latinx gang member who goes undercover as a police officer, DEA agents chase gang members through the streets of Boyle Heights (S1E7). Similarly, a Latinx gang member in *Mayans MC* (FX 2018–2023) cruises a Boyle Heights street looking for his mother, who works there as a prostitute (S1E1). Hollywood is particularly fond of shooting under the Fourth St. viaduct connecting Boyle Heights to downtown. *Goliath* (Amazon, 2016–2021) and *The Rookie* (ABC, 2018–) use the same blue house under the bridge for violent crime scenes (S2E1, E2; S1E1). A three-minute drive away, the iconic Mariachi Plaza is also a popular filming location. In the first episode of *Bosch* (Amazon, 2014–2021), the LAPD homicide officer of the title tracks a Latinx suspect from the Alameda Street metro station to Mariachi Plaza. *Goliath* introduces viewers to Los Angeles mayoral candidate Marisol Silva at her press conference in the Plaza, the statue of Mexican ranchera singer and Golden Age film performer Lucha Reyes visible behind her as Silva denounces a Mexican cartel for which she secretly works (S2E1). In the second episode we see her walk from the metro station through the Plaza, where the corrupt Mexican politician briefly sings "Volver" with two mariachis before heading to her nearby campaign headquarters (S2E2). Finally, episode 3 uses the plaza to stage a nighttime carnival where Silva and the crusading lawyer Billy McBride (Billy Bob Thornton) meet as mariachi music plays in the background.

Mariachi Plaza is a rare memorial to vernacular Mexican culture in Los Angeles. As Pulido, Barraclough, and Cheng write in *A People's Guide to Los Angeles*,

> Consumers of mariachi music have long known that they can hire musicians at Mariachi Plaza, making this site one of the older "shape-ups" in Los Angeles. A shape-up is a site where workers in the informal economy congregate while awaiting work. . . . The growth of the informal sector is part of the region's increasingly polarized economy, which is characterized by both high-wage and low-wage employment. Though mariachi musicians are not usually associated with the day laborers who provoke ire and angst among some segments of the population, they too are caught in the more vulnerable end of this polarized economy as it manifests in Los Angeles.[28]

One consequence of the polarized economy in the neighborhood is the residential displacement of mariachis and other day laborers due to gentrification.[29] Many shows shot there, however, washout gentrification with representations of Latinx criminality, seemingly justifying the increased police presence in Boyle Heights, particularly to protect new businesses against protesters.

Gentrification and the Police in *The Walking Dead*

The Walking Dead makes explicit the carceral ideologies and practices implied by gentrifying, mixed use studio housing developments. The series is often about efforts to redevelop human civilization, represented by the walled and fortified communities of Woodbury, Hilltop, Alexandria, Sanctuary, and the Commonwealth, which were all filmed in Senoia, Georgia. *The Walking Dead* projects a Hobbesian world where zombies are dangerous but the living are worse, justifying all manner of preemptive violence in the name of survival.[30] As Lauren O'Mahone, Melissa Merchant, and Simon Order argue, the show represents a necropolitical world where its central protagonist, a deputy sheriff from a rural town on the outskirts of Atlanta named Rick Grimes, exercises authority by making brutal life-and-death choices.[31] He also constantly bellows about "protecting my family" at any cost, thus inadvertently suggesting the kinds of violence disavowed in the Senoia development's invocation of "family values." The real Senoia housing development was the location for the fictional community of Alexandria, a former upscale "green" suburban development led by a Hillary Clinton-like Ohio Congressional representative named Deanna Monroe who agrees to provide sanctuary to Grimes and his group. The latter think the inhabitants of Alexandria are naive about and ill-prepared for zombie and human threats, warning them about the need for regular patrols, a task which Grimes and his partner take on themselves, even donning makeshift uniforms. Their insistence on vigilant security measures proves prophetic when the community is attacked by both zombies and the living, and many of the "soft" inhabitants of Alexandria die because they don't heed Deputy Grimes's warnings, including Monroe. As a kind of double for studio housing developments, Alexandria and other fortified communities in *The Walking Dead* suggest that TV industries are partly anchored in patriarchal ideas about safety and security.[32] More precisely, Alexandria was apparently conceived as an actual housing development. Many of the sets for the community were constructed as real, functioning homes, which were subsequently sold, as previously mentioned.

In contrast with Alexandria, which central characters romanticize as "home," the rival community called the Commonwealth represents the most dystopian interpretation of studio housing developments. It is a prosperous community with restaurants, stores, and theaters that looks like redeveloped Senoia (even though shot on backlot sets nearby). Suggesting comparisons to the actual gentrifying housing developments connected to light rail routes around Atlanta, the Commonwealth is centered on a train station that connects it to other, post-apocalypse townships. With the aid of an army dressed in storm-trooper-like armor, Commonwealth elites control the mass of poor people in the settlement through threats of expulsion and forced labor on the railroad. Meanwhile, within Commonwealth's fortified gates is a second gated community called "The Estates," where wealthy survivors are protected from both zombies and their working-class neighbors. The Commonwealth thus seems like a gothic rendering of *The Walking Dead*'s conditions of production within local racial capitalism. The show screens like a symbolic rendering of the marriage between gentrification and the police.

Throughout *Producing Precarity*, I have focused critical attention on the practical, working alliance between TV producers and the police, including the employment of police as actors and consultants; Hollywood's subsidization of policing in the form of equipment and location fees; and their shared interests in the occupation of Indigenous land, including Georgia's Shadowbox/Blackhall Studios and the controversial police training center "Cop City." This alliance is reinforced by TV show representations of police departments as diverse institutions devoted to serving diversity. The shows in this chapter emerge from and mediate local contexts where the gentrification produced by studios and their linked housing extends Hollywood's support for the police to protect private property. Translating its surrounding conditions of production into settings for post-apocalyptic horror, *The Walking Dead*, for example, indirectly suggests that police power is constitutive of gentrifying mixed use studio developments like at Senoia.

The Curse in New Mexico, or the White Saviors of Gentrification

The Curse is about a white newlywed couple, Whitney and Asher Siegal (Emma Stone and Nathan Fielder), who are building an eco-conscious housing development in Española, New Mexico, while filming an HGTV reality series about their efforts titled *Flipanthropy*, directed by their friend, Dougie (Benny Safdie). *The Curse* is set where it was also filmed, in the actual, working-class town of Española, which is 84% Latinx and about 2% Native American, with 19.9% of its population living below the poverty line (compared with 11.1% nationally). *The Curse* begins with the couple filming a scene in the apartment of a working-class Mexican man named Fernando, who sits on a couch next to his unnamed mother. He proceeds to tell the couple and their crew about his difficulties finding work to support himself and his mother, who has cancer. Whitney and Asher excitedly tell him that they have a job for him as a barista in a new cafe they've opened in town. (As if speaking to the overblown promises of TV jobs, it's later revealed that the café and hence the barista position are only temporary, for the duration of the filming of the show.) When the director doesn't find the mother's reaction to the job news dramatic enough, he insists on pouring water on her face to simulate tears.

The Curse satirizes racist condescension by reproducing it in the production of the show within the show. The mother has no name or lines and the actor who plays the part is not listed in the show's credits. Rendered mute and nameless, they were subjected to racist, sexist treatment as part of their role in the satire. This exemplifies Cedric Robinson's suggestion that in early twentieth century Hollywood, extras of color were often required to perform in demeaning and dangerous racial spectacles (chapter 1). In practical terms, there is little difference between the show's satire of racial capitalism and the operations of racial capitalism itself. *The Curse*'s producers take advantage of the town of Española to make a show about TV producers who take advantage of Española.

A satire about the building of expensive eco-homes, *The Curse* takes advantage of state subsidies, spending money that could go to low-income housing in Española. At the same time, the show took advantage of public resources for filming locations. Episode 6, "The Fire Burns On," was largely filmed at the Española Fire Department, with city fireman playing themselves. As a state-owned building, the fire station would have been available to *The Curse* for a modest fee, which amounts to another subsidy since it would be expensive to build a fire department set dressed with expensive equipment. Finally, *The Curse* took advantage of the unpaid labor of local people. Showrunner, co-writer and co-star Benny Safdie explains, "as we were looking to find the Cara character, we interviewed a lot of different Native artists from the area, asking about the struggles they had to deal with and the people they ran into. All of that went into the show. It was all in service of making it feel as real as possible." The character Cara Durand (Nizhonniya Austin, Diné and

Tlingit) is an Indigenous artist who Whitney tries to convince to lend her work to decorate a model home on *Flipanthropy*. Whitney projects respect and reverence for Indigenous people as she fetishizes and appropriates Indigenous art. Yet Safdie's sentiments seem similar. He voices respect for Indigenous "struggles" and idealizes the reality they represent, while at the same time describing how he wrote the script by appropriating the unpaid labor of the artists he interviewed. Whereas Whitney ultimately pays Cara for serving as a consultant on *Flipanthropy*, the credits for *The Curse* do not include any Indigenous consultants.

Similarly, by casting local actors in small roles, the program's producers were able to appropriate the unpaid labor of representing authenticity. As Safdie explains, "we wanted people from the area because we were going to learn from them. We would put them in the scene and then have conversations with them, saying, what do you do here? What are the biggest problems you have? Then we'd rewrite and put it in the show."[33] These were of course only temporary jobs, like those generated by Asher and Whitney in the show, and although local people were paid for their screen performances they were not compensated for their contributions to the script. Elsewhere Safdie has said of filming in Española, "We wanted to incorporate the entire community and all of their issues and problems into the show.... Basically, we just wanted to open up the world to everybody there because they know more about the community than we do.... You just have to be open to listening to them."[34] Here Safdie describes the appropriation of uncompensated contributions to the script in terms of liberal inclusion. Casting locals incorporated the community into the show, opening the world of TV production to them. At the same time, ironically, he seemingly compliments himself for being "open to listening to them" while talking about how he helped make *The Curse*. Safdie sounds like the hypocritical liberal do-gooders the show satirizes. As I've argued throughout this book, TV shows about the making of TV shows are often disavowed representations of the process of producing TV under local conditions of racial capitalism which nonetheless invite viewers to think in meta ways about the relationship between fictional TV shows and real ones. Here I would suggest that the depiction of the making of the fictional HGTV show *Flipanthropy*, with its racism and exploitation, performs the racial capitalism in *The Curse*'s mode of production.

Boyle Heights TV and Anti-gentrification Protests

Self-consciously departing from conventional TV representations of the area as a setting for crime, a trio of shows instead represent a queer Latinx Boyle Heights struggling with gentrification. The first is *Falling for Angels* (Here TV, 2017), an anthology show about gay romance in different parts of Los Angeles. The first episode, "Boyle Heights," written and directed by Latinx creative Nick Oceano, focuses on an assimilated, upwardly mobile Latinx man from Texas named Jesse (Luis Jose Lopez) who, despite his concerns about crime, agrees to move to Boyle Heights with his Anglo boyfriend Steven (Steve Grand). The episode opens with a shot of the Fourth St. viaduct followed by a montage that includes Mariachi Plaza, iconic Chicanx murals, and new art galleries. The montage is accompanied by a rap song about gentrification with the lyrics "displacement, dislocation, urban renewal restoration, Boyle Heights is for sale, highest bidders make your bets. . . . Stand up my *gente*, it's time to defend."

One morning, on Cesar Chavez Avenue, Jesse meets a young poet and activist named Leo (Adrián Núñez), who is handing out flyers calling for the boycott of a gentrifying coffee shop. The location is Weird Wave Coffee, an actual shop that has been the subject of local anti-gentrification protests. The two characters argue over the impact of gentrification, but Jesse ultimately invites him to a party at his parent's home. They flirt, and Jesse confesses that in response to racism "I guess I've been cutting myself off from something." The implication is that his encounter with Leo reconnects him to his Latinx identity. The two make out in the activist's room (decorated with a copy of Michel Foucault's *Discipline and Punish*) but Jesse stops short of having sex, choosing to instead return to his white boyfriend. Their relationship has been strained by Jesse's prior affair, but he makes an emotional apology and the couple embrace in a passionate kiss before heading to the bedroom. In the morning Jesse is more tolerant, claiming to like the loud hip hop played by his Mexican neighbors that he had previously complained about, and his love and desire for Steven are renewed.

While benefiting from government subsidies that divert money from social spending and promote gentrification, *Falling for Angeles* represents "*genteficacion*"—Latinx on Latinx gentrification—in a positive light. The show suggests that gentrification enables Jesse to renovate his Latinx identity while rekindling his relationship with his white romantic partner. *Falling for Angels* ultimately makes gentrification in Boyle Heights seem progressive, even romantic.

One year later, in 2018, the queer dramedy set in gentrifying Boyle Heights titled *Vida* streamed on Starz. The show is about sisters Emma and Lyn Hernandez (Mishel Prada and Melissa Barrera), who return to Boyle Heights after years away when their mother Vidalia (Rose Portillo) dies and leaves them her old-school lesbian bar. They learn that their mother was queer and married to a Chicana[35] named Edy (Ser Anzoategui), to whom Vidalia also willed part of the bar. *Vida* was created by queer Chicanx playwright Tanya Saracho, who is also the showrunner, writer, and the director of several episodes. Its other writers are mostly queer women of color, and Saracho self-consciously foregrounds a queer Chicanx gaze. This is evident in the show's carefully choreographed sex scenes, non-binary actors and characters, and gender-queer discourses, self-presentations, and practices, including a gay wedding where the couple are dressed like *norteño* musicians, and an elaborate queer *quinceañera* for one of Lyn's friends (S2E3, S3E4). Steeped in both *jotería* and *brujería*, as well as code-switching and Spanglish shade, *Vida* is perhaps the queerest show ever made about a Chicanx community. Episodes present critical representations of masculinity, Chicanx identification with and desire for whiteness, gender/sex policing in queer of color communities, and differences of generation and class (both within Boyle Heights and between West and East Los Angeles). Finally, *Vida* dramatizes the recent history of protests over gentrification in Boyle Heights, where it was partly filmed.

Inspired by a *Los Angeles Times* story about anti-gentrification protests in the neighborhood, an independent film production company commissioned a "world building document" about it from queer Latinx writer Richard Villegas Jr., which they then used to pitch a show idea to STARZ. An executive for STARZ then pitched to Saracho several Latinx properties, including a telenovela remake, a Santería zombie show, and *Vida*. STARZ told Saracho they wanted a program set in East Los Angeles, featuring millennial women, about "chipsters" (Chicanx hipsters) and "*gente*fication." Saracho says she added queerness to the mix, but it's clear that the East LA *gente*fication narrative was a corporate, algorithmic choice aimed at appealing to a particular demographic of Chicanx viewers.[36]

From its inception, *Vida* has focused on brown on brown "*gentefication*" within Boyle Heights to the relative exclusion of gentrification from the wealthier and whiter west side of Los Angeles. While local protesters have, as depicted in the show, opposed *gente*fication, their activism has been even more focused on economic redevelopment projects from outside the neighborhood, notably by the United Talent Agency. Headquartered in Beverly Hills, UTA has built an art gallery in Boyle Heights promoting artists from outside the neighborhood. Local protestors have posted mock eviction notices on the UTA gallery and marched behind banners reading "Keep Beverly Hills Out of Boyle Heights." UTA is also the agency that represents Saracho, and *Vida* was itself the object of anti-gentrification protests, forcing the show to limit location shooting in the neighborhood. As a result, many scenes were filmed in the Pico Union neighborhood of Los Angeles, home to large numbers of working-class Mexican and Central American people.

Anti-gentrification activism is represented in *Vida* by the Vigilantes, which Anastasia Baginski and Chris Malcolm argue represent "a made-for-TV composite of community resistance groups in East Los Angeles," including BHAAD. The organization protested location shooting for *Vida*, which benefited from state tax incentives. Among other actions, the organization attempted to block filming in Boyle Heights' Mariachi Plaza, which is featured in the series finale.[37] In their protest against *Vida* filming in Mariachi Plaza, BHAAD argued that the show had appropriated and belittled its members' images (including the members of the Ovarian Psycos, a feminist bike patrol which serves as one model for the Vigilantes) and contributed to gentrification.

One of *Vida*'s central characters, Mari Sanchez (Chelsea Rendon), is a bike-riding member of the Vigilantes who joins them in agitating against the Hernandez sisters and their efforts to "update" their mother's bar and attract younger, wealthier customers. Mari's militant attitude toward Lyn and Emma, who she derisively calls "coconuts" and "whitinas," softens when Emma gives her a job and a place to stay after she's kicked out of her home by her father. In the season 2 finale, she warns Emma that the Vigilantes are planning to disrupt the bar's new chipster music night, but in the ensuing protest Mari stands nervously on the sidelines,

Figure 7.4: "Wash Her Out!" Anti-gentrification activist Yoli bombs Lynn Hernandez with laundry detergent.

ignoring her best friend Yoli's entreaties to join in. Meanwhile, Lyn makes an impassioned speech to the protesters, claiming a genealogical link to the bar, which was built by her grandfather and owned by her mother, and reminding them that she and her sister grew up in Boyle Heights. Unmoved, Yoli throws powdered white detergent in her face as protesters chant "wash her out!" (See Figure 7.4.)

Emma then runs out of the bar, tackles the activist, and punches her in the face before friends pull her away. As Emma cradles her sister, the police arrest Yoli, who we subsequently learn is a DACA student vulnerable to deportation.

Vida disappears state-sponsored redevelopment schemes to focus instead on individual actors and choices. The second season finale stacks the deck against the anti-gentrification movement. Mari's reluctance to join the protest, Lyn's emotional speech, Emma's protection of her sister, and, finally, Lyn's vulnerability (the soap burns her eyes and her sister must support her while walking to the bar's bathroom) mobilize sympathy for the gentrifiers over and against the protesters.[38] This opposition is reinforced visually, in terms of hegemonic beauty norms, since Emma and Lyn look like tall, thin dress-wearing models while Yoli is shorter, rounder, and costumed in less femme and more functional jeans and plaid shirt. Finally, although the scene correctly suggests that gentrification has brought a greater police presence to Boyle Heights, *Vida* seems to affirm Emma's attack on Yoli and the activist's subsequent arrest.[39] In the show's third and final season, Mari appears to follow the sisters' lead. She grows out of the Vigilante's militant group think, resolving to "go it alone" by providing content for a socially conscious digital magazine (S3E6). Boyle Heights' anti-gentrification movement is thus caricatured as childish and ineffective, so focused on trivial challenges to rigid ideas about authenticity that they are unable to recognize "real shit" like mass deportations, as if anti-gentrification activists aren't also engaged in immigrants' rights organizing. In these ways, *Vida* exemplifies what we might call "TV-washing," replete with bad-faith self-justifications for state-sponsored gentrification.

Shortly before *Vida* ended, the similarly themed show *Gentefied*, which also filmed exteriors in Boyle Heights, began streaming on Netflix. Partly to avoid protests, though, *Gentefied* was mostly shot on an enclosed soundstage at the newly established LA Hangar Studios in an industrial section near Boyle Heights, next to a factory that manufactures coffins.[40] While more modest than developments in Georgia and New Mexico, the studio is still a force of gentrification in the area, a sort of adjunct to Boyle Heights and the nearby LA arts district.

Gentefied centers efforts by the Morales family to keep their beloved taco shop open. "Mama Fina's" is named for the late wife of Casimiro Morales (Joaquín Cosio), who owns the shop and employs his three grandchildren, Ana, Erik, and Chris (Karrie Martín Lachney, J. J. Soria, and Carlos Santos). Like *Falling for Angels* and *Vida*, *Gentefied* is queer affirming, featuring Ana Morales and her Dominican girlfriend Yessika. The show also includes Black Latinx characters, such as Yessika and others, and it dramatizes anti-Blackness in Latinx communities.

As the title suggests, however, gentrification is the program's central theme. In the first episode, Casimiro is behind on his rent, and his smarmy landlord, Roberto (Wilmer Valderrama), brings two potential buyers to Mama Fina's taco shop. As he enters the shop, Roberto says "imagine owning in Eco Park five years ago. That's what Boyle Heights is." Casimiro and his grandson Erik rush from behind the counter to confront them, and Roberto assures the prospective buyers, who are taking pictures on their cell phones, "They're harmless. . . . Get some tacos."

Her family facing eviction, Ana, who is an artist, participates in a group show at a local art gallery with an anti-gentrification collage made from yellow and green signs with black letters declaring "WE BUY HOUSES."

Over time and several episodes, *Gentefied* represents the costs of gentrification. For example, Javier, the mariachi who bitterly complains about Mama Fina's rising prices in episodes 1 and 5, faces displacement in episode 6, and is forced to leave the neighborhood for Bakersfield.[41] At the end of season 1, episode 1 of *Gentefied*, Casimiro is arrested by the police for throwing a bottle at a gentrifying development, while in the season 1 finale, he is arrested by ICE, thereby connecting gentrification and detention. With the drama of Casimiro's possible deportation in season 2, *Gentefied* presents compelling, critical representations of the MAGA-era deportation regime and how it complements gentrification. After receiving media training by immigrant rights activists, Casimiro appears on a TV news show hosted by CNN contributor Van Jones, but Casimiro goes off script and criticizes the respectability politics informing media accounts of worthy versus unworthy migrants.

> I'm tired of sharing all the good things I have done in this country, to make you feel bad for me? To convince you that I am a person? Look at me. I am a person. I have done good things. I have done bad things like everyone. (In Spanish) And what? You want to . . . Because I don't have a piece of paper you want to destroy my family? (In English) Why? Why would this country do that to millions of people? To be treated like animals. (In Spanish) Not even that. Even dogs are treated better than an undocumented person. (In English) This country promised the American dream, but you lied. I work hard. (In Spanish) But you want to rip the dream from me (in English) I'm done begging. I'm . . . I'm not a perfect immigrant. But I know I am not a criminal. (In Spanish) Stop lying (In English) This is my home. This is where I belong, whether you like it or not.

Casimiro refuses a respectability framework for distinguishing between good and bad immigrants. He does so by looking into the camera and directly addressing viewers, both the viewers of the TV show within the show and the viewers of *Gentefied*. Casimiro thereby breaks the fourth wall to reject the building of walls at the border.

As this scene indicates, like so many others in this study, *Gentefied* is a TV show partly about TV shows in ways that invite interpretations of it as a self-reflexive reflection, disavowed or not, on what it means to make state-subsidized shows in poor places. Such reflection is suggested by the episode titled "Protest Tacos," where the Morales family attempts to counter anti-gentrification protests by filming their own viral commercial. Under the threat of eviction, the family decides to self-gentrify by reducing portions and using fancy new ingredients and preparations. At the start of the episode, Erik passes out yellow flyers for a food tour that the taco shop is part of called "Bite into Boyle Heights." The yellow flyers with black lettering recall the yellow and black "WE BUY HOUSES" signs in Ana's anti-gentrification collage, suggesting the taco shop's complicity in gentrification. Yessika confronts Chris about the tour, arguing that "welcoming outsiders en mass with open arms like this is pushing people out of their homes and into the tents around every corner." When Chris protests that they need the money, she responds, "You think there's only one way to help save the shop? That your only option is selling out your community, my community?" As she leaves, she threatens to protest.

Ultimately, however, Mama Fina blunts the effect of negative publicity on their bottom line by incorporating protests into their advertising. As Erik asks, how can they convince hipsters from outside the neighborhood to cross a picket line? Answering his own question, he says "Let's lean into the spectacle of it all" by making it seem like the anti-gentrification protest is part of the tour. Comparing it to the Teatro Campesino skits in support of striking United Farm Workers, Erik directs a satiric video with Chris playing a gay gentrifier who encounters cartoonish protestors in front of Mama Fina's, carrying signs reading "Stay on the West Side" and "Down With Whitey," while derisively calling Chris "Christopher Columbus!" Next Erik, dressed as a cholo, comes out of the shop and faces the camera, broadly smiling, and says "This Sunday, come take a bite of Boyle Heights, before it's gone forever." Chris ends the video by telling viewers that at Mama Fina's, they can both protest *and* eat tacos. When the actual protestors arrive, one is carrying the exact same "Stay on the West Side" sign as the fake protestors in the video. The advertising video works, however, and a line of mostly white people file into the shop, with one woman saying, "This is like so cool, we can get tacos and make a difference." Ironically, Erik and the others borrow from a farm workers' union to advertise their business. This is a fictional version of the production of *Gentefied*, which appropriates forms of protest art to make and advertise a commercial product.

The program repeatedly stages debates over the role of art in gentrification. In an episode titled "The Mural," for example, Ana's white patron, Tim, has recently bought a building in Boyle Heights which will house a gallery that he promises will bring together artists from East Los Angeles, West Hollywood, and Santa Monica, and he commissions her to paint a mural. Standing in front of the blank wall he says "And where the artists go, the money goes. Sí se puede!" However, a long-time tenant named Ofelia, who runs a liquor store in part of the building, objects to Ana's mural. Titled "Brown Love," it depicts two Latinx men wearing lucha libre masks pulled up over their mouths and passionately kissing. To represent the community response, the show uses a series of low-angle shots depicting different characters—Javier and his son Danny, two elderly customers from Mama Fina's, two young women, and two cholos—as if in POV shots from the perspective of the art. Ofelia is furious that the mural is alienating her regular customers, driving them to 7-11, and in the end the mural is destroyed by taggers. But before that, *Gentefied* dramatizes a debate about gentrification and art among the Morales cousins, Yessika, and her friend as they eat bacon-wrapped Mexican hotdogs. Yessika summarizes the debate by saying both sides have a point. "Yes, it was commissioned by a fucking colonizer, but I'm all for fighting *comadre* homophobia with queer love bombs." This line can be read as a version of the contradictions represented by *Gentefied*, in which a gentrifying Netflix commissions progressive art about Latinx people.

The contradictions between the TV show's content and its corporate context are suggested by the Netflix art produced to advertise *Gentefied*. Historically, murals in Chicanx communities have been linked to small business and often represent history and politics. Netflix appropriated the mural form for advertisements featuring the four central characters depicted on a mural on the side of a brick building. On the bottom corner of the mural are the torn remnants of a sign for "New Luxury Apartments," like one of the signs in Ana's anti-gentrification collage. The mural is captioned "THIS FAMILY'S NOT FOR SALE" but is part of an advertisement selling a TV show about them and, more broadly, about gentrifying Boyle Heights. It's not that Netflix buys such people and places, but it does mine Boyle Heights for profits. With its dependence on low-wage labor and poor locations, TV production in Boyle Heights is an extractive industry.

This is the contradictory TV world that many Latinx creatives navigate. *Gentefied* was created by Linda Yvette Chávez and Marvin Lemus. Raised by Mexican immigrant parents in Norwalk, California, Chávez studied with Chicana feminist writer Cherríe Moraga while an undergraduate at Stanford before earning an MA in screenwriting from USC. Lemus, with a Guatemalan father and Mexican mother, was born and raised in Bakersfield, and graduated from the Los Angeles Art Institute. The trajectory of the displaced mariachi character Javier is the inverse of Lemus's, who moved from Bakersfield to Boyle Heights in 2014, six years before *Gentefied* first streamed, and drew on his experience as a new resident in the gentrifying neighborhood while writing the show. As he told a reporter, "I was part of the change coming into the neighborhood, even though I grew up in a neighborhood just like this. And it was devastating." Chávez and Lemus brainstormed for *Gentefied* at Boyle Heights Café, which subsequently lost its lease and was forced to relocate.[42] Like *Vida*'s Saracho, the young Latinx writers played a part in TV gentrification, but a modest one relative to corporate giants Netflix and Starz. Indeed, *Vida* and *Gentefied* were both prematurely cancelled, speaking to the ongoing disposability of Latinx shows and talent.

Comparative Settler Colonialisms

For all it obscures, *The Curse* is a revealing representation of TV production as a form of settler colonialism. While making *Flipanthropy*, potential buyers for one of Whitney and Asher's houses back out over the threat of Indigenous land claims. In addition to the Indigenous artist Cara and the plot about the appropriation of her art, the show also includes the character James Toledo (Gary Farmer, Cayuga Nation and Wolf Clan of the Haudenosaunee/Iroquois Confederacy), Governor of the fictional San Pedro Pueblo, who talks with Whitney about the Pueblo land easements over which local highways pass. He joins the couple on the show within the show for an Indigenous land acknowledgement at the construction site of one of their houses (although notes from the network lead to his remarks being edited out). Speaking for him, Whitney looks into the camera and says, "Even though we all are occupying unceded Indigenous lands, the battle he speaks of is less a literal battle and more of a clerical concern over compensation for road use in areas that don't affect access to our properties, or anything related to titles of this land, even though we do recognize ourselves as having obtained this land through a long history of colonization" (E9). While a parody of liberal white land acknowledgements as empty gestures, Whitney's speech also screens like an inadvertent satire of the show itself and, more broadly of TV production in New Mexico. Her remarks could easily be about Netflix studios and the connected Mesa del Sol housing development in Albuquerque, which occupy Indigenous land in the heavily policed South Valley.

If a TV show about gentrification in New Mexico is a settler colonial project, what does that say about production in other places? The fact that in New Mexico—with its 19 Pueblos, three Apache tribes and the Navajo Nation—settler colonialism seemed like a striking and "live" critical issue to the producers of *The Curse* perhaps suggests that, by contrast, settler colonialism is generally occluded in Southern California screen industries. Yet all the shows discussed in this chapter (and book) were filmed on Indigenous land.

As I noted in chapter 2, Georgia's Blackhall/Shadowbox studio and the adjacent Cop City police training facility occupy Muscogee (Creek) ancestral lands and have galvanized Muscogee (Creek) protests. Meanwhile, *The Walking Dead* survivor settlement "Terminus" is named for the first, post-removal white settlement in what is now Atlanta, while more broadly the show's Hobbesian view of the war of all against all is definitive of frontier and settler colonialism.[43] In California, the entire Los Angeles basin is Tongva land. Which means that all of the major film and TV studios (Warner Brothers, Paramount, NBC Universal, Walt Disney, and Sony) as well as screen institutions (the UCLA and USC film schools, the Academy of Motion Picture Arts and Sciences, the Paley Center for Media, the Directors Guild of America, and the Writers Guild of America), wealthy film industry residential communities (Beverly Hills, Bel Aire, Century City, Studio City), and iconic film locations (Sunset Boulevard, Griffith Observatory, City Hall, the Los

Angeles River, the Bradbury Building, Union Station, the Santa Monica Pier, even the Brady Bunch house) all occupy Tongva land. Finally, gentrifying Boyle Heights is on Tongva land. From the perspective of Tongva land claims, among other things gentrification is part of the history of settler colonialism in Hollywood and beyond.

The situation comedy *Rutherford Falls* is the rare show that reminds viewers of Hollywood settler colonialism. Sierra Teller Ornales (Navaho, Edgewater clan, born for the Mexican people) was its co-creator, writer, executive producer and showrunner, and half of *Rutherford Falls*' writers were Indigenous, including the program's co-star, Jana Schmieding (Cheyenne River Lakota). The town of Rutherford Falls sits on the historic land of the fictional "Minishonka" people. The first shot of the first episode is a closeup of a bronze statue's face, a man in a seventeenth century Puritan hat. As the camera slowly moves from closeup to a long shot, the figure is revealed standing on top of a large pedestal. Next, we see a red car come crashing into the statue, which is that of Rutherford Falls founder Lawrence Rutherford in the town's historic square. (See Figure 7.5.)

His descendant, Nathan Rutherford (Ed Helms), runs the local historical society out of his ancestors' home (an early form of settler gentrification), and he is committed to preserving the memorial.

Figure 7.5: A settler colonial memorial on Universal's Court House Square in *Rutherford Falls*.

But after the accident, only the most recent of many, Mayor Diedre Chisenhall proposes moving the settler colonial memorial. When he protests, the mayor says, "Nathan, in case you haven't noticed, this isn't a great time for people who love statues." Here the mayor implicitly refers to recent protests over a statue of Robert E. Lee in Charlottesville, Virginia (chapter 6). Indeed, the statue resembles a Confederate memorial, but with Rutherford holding a Bible instead of a rifle. It sits in view of the town's City Hall, like so many Confederate monuments in front of courthouses to project state support for white nationalism. But instead of Confederate flags, during the celebration of "Founder's Day" the settler colonial memorial is surrounded by US flags and red, white, and blue bunting and balloons, emphasizing the articulation of settler colonialism to white nationalism. Nathan's dogged insistence on keeping the statue in the town square puts him at odds with long-time Minishonka friend Reagan Wells (Schmieding), the curator at the Minishonka cultural center, who calls attention to the irony of a white settler bemoaning the loss of his history. When Nathan says, "that statue is core to who I am and if it can just be picked up and tossed around willy-nilly with no regard for its historical specificity then literally, who am I?" she responds "Okay, what you're describing is my entire life. It's literally something I have to deal with every day" (S1E1).

Nathan makes a speech in the town square in front of city hall, where he shouts at opponents of the statue "to shut your stupid pie hole. . . . That statue, which sits on my family's land, commemorated all he gave us, and if you don't get that, well you're just an ungrateful boob. . . . If you disagree, you can burn in hell." When his rant goes viral it creates a PR problem for the Manhattan-based Rutherford Corporation, leading it to threaten to seize Nathan's land and ancestral home, since he is only an honorary board member of the company bearing his family name, with no real power. As part of a counterproposal to benefit the tribe and enable Nathan to keep his home and continue working at his heritage center, Reagan and Minishonka casino CEO Terry Thomas (Michael Greyeyes, Plains Cree) pitch to Nathan a mixed-use development called, in Reagan's words, "Ye Olde Rutherford Village, a cutting-edge retail tourism community," with period architecture and costumes, "like colonial Williamsburg." As Terry tells Nathan, "We're trying to make Disney World, and Nathan, we want you to be Micky Mouse." When

Nathan asks incredulously, "So you're going to take my land, but allow me to stay on it on your terms?" Terry answers, "Think of it as a fair and honest deal," repeating Nathan's claim about how his ancestor's acquired Minishonka land. In these ways, *Rutherford Falls* critically foregrounds settler colonial gentrification by inverting its conventional hierarchies between Indigenous peoples and settlers.

As part of their pitch, Reagan and Terry unveil an elaborate diorama incorporating the town square with the settler colonial monument into a tribal development. The diorama recalls a similar diorama in Nathan's heritage center. He even recognizes that the Minishonka diorama was made by the same artist. The town square depicted in the diorama represents one of the show's central filming locations, the familiar Courthouse Square at Universal Studios. These backlot sets have stood in for small white towns in *Back to the Future*, *Gremlins*, and many other works. The creators of *Rutherford Falls* thus appropriate the well-known location of white Americana to tell Indigenous stories about settler colonialism.

After seeing *Rutherford Falls*, it is difficult not to think of Courthouse Square as a settler colonial location when it appears in other films and TV shows. Ultimately, then, in a movement of meta reflection, the diorama of the proposed Minishonka development, like the show itself, imaginatively subsumes its backlot studio set within a larger, Indigenous space. This symbolic reversal mediates the actual conditions of production for *Rutherford Falls* at Universal Studios, which was built on Tongva land, at the former site of an Indigenous village called Cahuenga. While promoting the show, Schmieding reminded viewers that Southern California is Indigenous, wearing a baseball shirt resembling those worn by Dodger fans but instead reading "YOU ARE ON INDIAN LAND."[44] *Rutherford Falls* repeatedly references the Indigenous "Land Back" movement while symbolically performing it with its reappropriation of Universal's Courthouse Square.

In 2019, Tongva and Scottish artist Weshoyot Alvitre created a digital photocollage replacing the famous "HOLLYWOOD" sign (originally "HOLLYWOODLAND") with letters reading TONGVALAND. The image was subsequently used on two billboards (on Sunset Boulevard, in the gentrifying Silver Lake and Echo Park neighborhoods) and Alvitre circulated it widely on social media, including in this Tweet: "If you're watching the #Oscars2020: understand this: they are on Indigenous lands. They are on Tongva Lands. And We are still here. #decolonize #Hollywood #LosAngeles #Oscars."⁴⁵ Alvitre told *Indian Country Today* that with the TONGVALAND billboard they deliberately tried to "bring attention to the often-disregarded fact that the Hollywood hills, the city of lights, was built on unceded tribal lands." She concludes by effectively redefining gentrification as settler colonialism, recalling that "the original sign was an advertisement for new buildings. The sign we identify with lights, famous people and movies was made to advertise a new housing development to affluent white culture in 1923. The village of Kaweenga (Cahuenga) is in these hills."⁴⁶

In her art, Alvitre imaginatively reanimates the sign, making it flash between an ad for a white housing development and a "land back" demand. The Hollywood sign both occupies Tongva land while representing the larger network of screen industries in the Los Angeles Basin. As Alvitre glossed her TONGVALAND photocollage on Instagram:

> Never forget that when you are visiting Griffith Park, to view the Hollywood sign, you are on Tongva Land. Never forget that the most recognizable sign in the entertainment industry, and the industry itself in Los Angeles and Hollywood and Burbank is directly on top of Tongva Lands. Never forget the Hollywood stars celebrating every man and woman who had a successful career entertaining the world, and every Oscar statue, profited off of Tongva Lands. Never forget the hundreds of films, shot on unceded lands and sacred sites, forever memorialized in sandstone backdrops, mission remains, and sepia toned chaparral are on Tongva Lands. Never forget that even today, soundstages and locations for off world shows and financial successes of entertainment moguls and their streaming services lay on Tongva Lands.⁴⁷

In similar ways we must also remember that studio soundstages, shooting locations, and streaming film and TV productions occupy and profit from Indigenous lands in Georgia and New Mexico.

Afterword

Streaming Poor Places

State-subsidized TV made in poor places, I have argued, exemplifies contemporary racial capitalism in both content and mode of production. With colonial and plantation origins, racial capitalism in Georgia, New Mexico, and other poor locations has been translated into film and TV production. Promises of economic growth and especially employment are greatly inflated, while the local benefits of state subsidies remain small compared to the social costs of diverting public resources to Hollywood. TV production employs some local workers directly in production but significantly depends on racialized and gendered low-wage labor in the surrounding economy. New filming studios and their mixed-use housing developments result in higher housing costs that displace poor people and encourage increased policing to protect private property. Profiting at the expense of Indigenous people and people of color, TV is implicated in the violence of social disinvestment, gentrification, and police and prisons. TV studios have also become prominent leaders in the ongoing history of settler colonialism, combining gentrification and policing on Indigenous lands. In such contexts, many shows project progressive race politics that serve to displace from view and critical reflection the conditions that make such representations possible. DEI representations can thus serve a repressive end, as suggested by the many shows that depict the police as champions of diversity, equity, and inclusion. But as Pulido, Laura Barraclough, and Wendy Cheng argued at the outset, state and capitalist domination of different locations is "occasionally, and sometimes successfully, resisted by people, with an alternative vision of how the world should work." Examples of resistance to TV racial capitalism include the twin movement against Cop City and Blackhall/Shadowbox Studios' occupation of forests and waterways in Georgia, and the anti-gentrification movement in Boyle

Heights. As these examples remind us, studying TV filming locations can open critical perspectives on how poor Black and Latinx people oppose dominant constructions of poor places.

The largest number of shows analyzed in this book stream on digital platforms, with Netflix accounting for most. But shows on legacy and cable networks are also available for streaming. This has important implications for contemporary TV's spatial practices in poor places. Netflix and other platforms promote streaming as an advance in consumer choice. Streaming, the story goes, frees audiences from the tyranny of fixed schedules, enabling them to tailor the viewing experience to their own needs and desires. These freedoms are limited, however, by corporate algorithms that direct viewers to categories and genres that tend to reproduce status quo viewing practices. As Sarah Arnold argues, algorithms "use data gleaned from online user interactions as a way of profiling and controlling" audience behavior. While the personalized schedules made possible via streaming "might allude to the liberation of the individual from the mass" they also mask "more profound forms of individual manipulation and governance manufactured through data algorithms."[1]

Algorithms encourage viewers to stay in their comfort zones and watch more shows like the ones they already watch, which effectively reproduces white TV tastes and hence a white demographic. According to Mark D. Pepper, the "algorithm's suggestions effectively shape our sense of what matters into a self-gratifying mirror of previously validated ideas, tropes, and identities." He suggests that streaming algorithms result in de facto segregation, directing white viewers toward conservative white family sitcoms such as Tim Allen's *Last Man Standing* (described on Imdb.com as about "a married father of three (who) tries to maintain his manliness in a world increasingly dominated by women"), and away from family of color sitcoms like *Fresh Off the Boat* and *Black-ish*. Or TV algorithms may direct white viewers to what Jorie Lagerwey and Taylor Nygaard call "Horrible White People" shows, bleak comedies made after 2016 and the election of Donald Trump that center precarious middle-class liberal white people, particularly women. Lagerwey and Nygaard argue that HWP programs sync up with the MAGA era since, in both, audiences are confronted "with the failure of their white middle-class identity to grant them the privilege of stable or easy-to-find jobs, accessible home ownership, and long-term relationships."[2] As they write in *Horrible White People: Gender, Genre, and Television's Precarious Whiteness*, "the proliferation of Horrible White People shows, along with the industry that produces them, is complicit in the rhetorical shift toward White suffering that has helped sustain structural White supremacy and worked to support the rise of the political far-right in the United States and elsewhere where these shows are produced, distributed, and consumed."[3] They conclude that these so-called "quality" TV shows marginalize people of color while centering white vulnerability and resentment in ways that reinforce white supremacy.

Alternatively, streaming algorithms can hail Black viewers in problematic ways. As an example of group-specific digital marketing niches that respond to calls for diversity, Ruha Benjamin argues that Netflix recommendations targeting Black viewers may misleadingly foreground supporting cast members who are Black. "Why bother with broader structural changes in casting and media representation," Benjamin asks, "when marketing gurus can make Black actors appear more visible than they really are in the actual film? . . . (S)o long as algorithms become more tailored, the public will be given the illusion of progress."[4] With their racialized algorithms, moreover, streaming platforms reproduce a perceptual segregation for viewers that can reinforce segregation in the material world. The imagined freedom to choose a status quo of racial inequality via streaming thus complements the reproduction of racial inequality via state subsidies for TV location shooting. Building on Benjamin by revising the early term for Black and Mexican segregation, "Jim Crow/Juan Crow," we can analyze streaming as part of a new "Jim Code/Juan Code," or "the employment of new technologies that reflect and reproduce existing inequities but that are promoted and perceived as more objective or progressive than the discriminatory systems of a previous era."[5] Ideologies of racial progress via streaming thus complement progressive, DEI TV content.

Streaming platforms promote the pleasures of binge-watching to similar effect. TV scholars have extensively analyzed the psychology and phenomenology of binge-watching. Zachary Snider argues, for example, that binging produces intense "emotional self-immersion" and "empathic" involvement in TV worlds that he calls "narrative transportation." Binging "psychologically affects viewers' perceptions of reality" by imaginatively transporting them beyond the mundane every day in ways that can compromise their "real-world judgements."[6] Meanwhile, Mareike Jenner claims that Netflix encourages binge-watching as an "insulated flow" of viewing. The streaming platform discourages exiting the flow by, for example, automatically starting the next episode or reminding viewers of unfinished shows.[7] Although presented by corporations as new, binge-watching recalls classical film spectatorship. Elsewhere, drawing on Miriam Hansen's analysis of early film spectatorship, I have argued that silent film production and exhibition norms aimed to constitute respectable, middle-class white spectators focused attentively on a central, narratively coherent feature film—the binge-watching of the day. Over the course of the 1910s and 1920s, filmmakers strove to produce the "diegetic illusion" whereby the film fiction becomes its own, self-enclosed world. The suspension of disbelief necessary for the illusion encouraged spectators to take for granted that the fictional world of the film was perceptually segregated from the space of the theater, and more generally, from Southern California social space. Such spectatorship reinforced the imaginary autonomy of the fictional film world from its larger production and reception context. I argued that early Hollywood films transported white audiences beyond Mexican Los Angeles, psychically insulating them from the low-wage Mexican workers on which the film industry (and others) depended.[8] In similar fashion today, binge-watching transports viewers beyond their surrounding social spaces, including many poor places. The insulated flow of streaming platforms also insulates viewers from knowledge about TV's conditions of production. Binging promotes immersion in fictional settings at the expense of attending to real locations. Together, contemporary TV's modes of production and reception detach settings from locations such as Albuquerque, Atlanta, and Boyle Heights, making it more difficult to critically reflect on race and the politics of place.

By contrast, in this book I've tried to interrupt contemporary TV's insulated flows by reconnecting settings to material locations, highlighting collaborations between governments and media corporations and their responsibility for (re)producing inequality. The method of analyzing contemporary TV that I develop here could be called "streaming in reverse." Rather than focusing primarily on the final product as it appears on our screens, I've streamed TV shows backwards, as it were, asking where they all began. Rather than abstracting them from their shooting locations, streaming in reverse views TV programs concretely, as interventions into the politics of place. Streaming in reverse means to investigate the material conditions of possibility for TV shows in histories of white supremacy and resistance to it. Streaming in reverse thus makes visible the poverty that both attracts TV production and prompts popular protests.

The current and future transformations of TV production made possible by artificial intelligence represent possible limits to the method of streaming in reverse I have developed here, or at least suggests the necessity of its significant revision. The use of AI as a film- and TV-making tool promises to advance the logic of TV production in poor places while potentially moving beyond its contemporary local forms. Hollywood unions, notably the Writers Guild of America, have organized in favor of regulating AI to safeguard jobs for writers, actors, and others. Its status as an enemy of screen workers partly inspires films and TV shows, I would argue, about the dangers that AI poses to humans.[9]

The Georgia-shot show *Brockmire* (2017–2020), for instance, presents a bleak satire of a world dominated by AI, including television production. The title character is a grotesque, drunken baseball announcer who, over the course of four seasons, gets sober and becomes a better person while the near future world around him spirals into deepening economic inequality, climate change, and what the show's creator, George Church-Cooper, calls the dismantling of "the social safety net" and the domination of "a predatory capitalism that is picking off the poor and sick for spare parts."[10] *Brockmire* is thus a TV show about TV shows set in a dystopian future defined by state disinvestment in social spending, a future that seems extrapolated from *Brockmire*'s Georgia conditions of production. The suggestive similarities between the show's setting and its production context sheds critical light on the program's depiction of AI and its significance for the future of making TV in poor places.

As Brockmire becomes the commissioner of professional baseball, a TV spectacle in decline, a powerful artificial intelligence entity emerges. On opening day, 2031, Brockmire's daughter Beth gives him a gift, a small digital device in the shape of a smiling yellow lemon called "Limón." She explains that "it's this amazing digital assistant. . . . It's the algorithm. So if Google's like a chainsaw Limon's like a scalpel. It predicts all your needs, it adapts to your whims, and it learns at, like, an exponential rate." Her father jokingly responds, "smarter robots, now our dystopia is complete," which turns out to be prophetic (S4E2). Subsequent episodes depict the social diffusion of Limón in the form of lemon-shaped home devices and small lemon-shaped pins through which users can speak with artificial intelligence that mimics a human woman's voice, like Suri or Alexa. Brockmire develops an antagonistic relationship to Limón, freaked out by how it spies on him, and how it anticipates, cultivates, and channels his interests and desires. By 2034, Limón has destroyed Google, Facebook, and Apple, and it convinces Brockmire that it's *his* idea to meet and propose that the AI take over the management of baseball, including by streaming games and related content. The meeting is in a bare white room where Brockmire meets Limón's corporate board members dressed in lemon yellow and seated at a conference table. It's revealed that the Limón CEO is an AI that speaks through a white woman named Debbie. People are terrified of the AI so Limón uses Debbie as a mouthpiece because "empirically," she's "the sweetest person in America." Implicitly invoking baseball's historic anchoring of white, US nationalism, Limón effectively proposes to use the sports spectacle to prop up racial capitalism, or what it refers to as "market domination" via "opportunistic nationalism." The AI announces to the room that its "patriotic embrace of baseball" will pave the way for Limón-backed candidates who will "pursue policies of social justice reducing inequality and lifting any and all regulations on Limón." Finally, the AI predicts that its embrace of US nationalism in the form of streaming sports TV will help "defeat the evil Chinese Limón. There can only be one." To be sure, AI is and will be used in oppressive ways by corporations and state powers, including the police.[11] But the actual consequences of AI for streaming TV are not like *The Terminator* scenario depicted in *Brockmire*. Film and TV shows often represent the dangers of autonomous AI, but in terms of screenwriting it remains a tool and as of this writing not yet an independent generator of content.[12] AI nonetheless represents potentially dramatic transformations in the process of making TV shows in poor places.

The threat to jobs for writers and actors isn't a significant issue for places like Georgia and New Mexico, where Hollywood contributes few such jobs while supporting gentrification and low-wage racialized and gendered service work that reproduce and extend inequality. Instead, AI raises the prospect of Hollywood production physically withdrawing from the two states. In a 2024 interview with the *Hollywood Reporter*, Tyler Perry suggested that AI was a job killer, but in the present context his comments on its significance for the soundstages and locations of TV production are particularly striking. He announced he was putting on hold plans for an $800 million expansion of his studio to add 12 new soundstages because of a demonstration he witnessed of the Sora text-to-video OpenAI: "I no longer would have to travel to locations. If I wanted to be in the snow in Colorado, it's text. If I wanted to write a scene on the moon, it's text, and this AI can generate it like nothing. If I wanted to have two people in the living room in the mountains, I don't have to build a set in the mountains, I don't have to put a set on my lot. I can sit in an office and do this with a computer."[13] This is not the autonomous AI of science fiction, producing film and TV content on its own; Perry doesn't imagine putting himself out of a job as a writer but rather using AI to realize his scripts. While not perfect, the videos released in 2024 on the company's website show that Sora is particularly good at rendering cinematic locations such as busy urban streets and western towns and landscapes, potentially enabling TV producers to bypass location shooting and the building of sets.[14] By suggesting that AI may make studios and location shooting obsolete, Perry brings into sharp relief the question, what if Hollywood abandons places like Atlanta and Albuquerque? Both would lose an industry that supports policing and prisons and that degrades the environment. What new possibilities might open up in absence of state-subsidized film and TV? Without the fantasy of economic growth via film and TV production, how might state spending be reallocated? While we can hope for "soundstage to plowshare" futures, the historic competition to attract film and TV production suggests we may be on the verge of a new race to the bottom, where different states and countries compete against each other to attract AI servers. Commenting on the Perry story, AI marketing advisor and *Forbes* writer Elijah Clark advocates "tax incentives for entertainment and tech companies to collaborate (and) catalyze an AI creativity boom."[15] It's easy to imagine states that currently provide subsidies for conventional film production shifting to AI subsidies. Like studios, AI servers threaten the environment by consuming huge amounts of energy, water, and materials such as lithium and cobalt.[16] Like conventional TV producers, AI will no doubt also be very good at generating pro-police and -prison content based on the recycling of existing images. But whatever the future holds, I wager that TV will still depend on racial capitalism in place, even if in new and unforeseen ways.

ACKNOWLEDGMENTS

Thanks to Henry Jenkins and Karen Tongson, editors of the "Postmillennial Pop" book series, and to NYU Press Associate Director and Editor-in-Chief Eric Zinner, and Editorial Assistant Furqan Sayeed. I developed the ideas that would become *Producing Precarity: The Costs of Making TV in Poor Places* in three publications, and I'm grateful to the editors and reviewers of each. For "Precarious Locations: Streaming TV and Global Inequalities," thanks to the editors of *American Studies*, Sherry J. Tucker and Randal Maurice Jelks, as well as two anonymous reviewers. For "Contemporary Television and Racial Capitalism in Place," thanks to editors at *Jump Cut*, Julia Lesage and Michael Litwack, and an anonymous reader. Finally, for "'Plenty of Room to Swing a Rope': *Watchmen* and the Racial Politics of Place" in *After Midnight: Watchmen After Watchmen*, thanks to volume editor Drew Morton. I also presented early versions of this work at the Latin American Studies Program at Washington University; the Unit for Criticism and Interpretive Theory at the University of Illinois, Urbana-Champaign; and the Latinx Visions: Speculative Worlds in Latinx Art, Literature, and Performance, Department of Chicana/o Studies at the University of New Mexico; and I am grateful to my hosts and audience members. Graduate students in my 2024 UC San Diego Ethnic Studies course on methods for media studies read parts of the manuscript and helped me think about how to end it. Thanks to Cristal Alba, Isaias Rogel, Carl Schmitz, Citlally Solorzano, Phúc To, and Christie Yamasaki. Subsequently, Isaias took an independent study course with me on Latinx and Latin American horror, and our conversations sustained me through final revisions of chapter 4. I also thank three graduate students whose research and intellectual engagement have helped make this a better book, Stephanie Martínez, Naaila Mohammed, and Liliana Sampedro. I taught Ethnic Studies 101TV three times while researching and writing *Producing Precarity*, and I want to thank my amazing undergrads, who have taught me so much, especially Paige Love and Helen Zhu. I'm also grateful to my excellent graduate teaching assistants for the course, Burgundy Fletcher and Bettina Serna. A number of friends and colleagues read parts of the manuscript and gave me invaluable feedback, so thanks to Patrick Anderson, Chera Kee, Veronica Paredes, Roy Pérez, and Chris Perreira. Finally, I'm immensely grateful to Shelley Streeby, who read every word and gave me brilliant advice.

TELEVISION SHOWS CITED

A Man in Full (Netflix, 2024)
Ambitions (OWN, 2019)
Atlanta (FX, 2016–2022)
Being Mary Jane (OWN, 2013–2019)
Better Call Saul (AMC, 2015–2022)
Big Sky (ABC, 2020–2023)
Black Lightning (The CW, 2018–2021)
Bosch (Amazon, 2014–2021)
Breaking Bad (AMC, 2008–2013)
Briarpatch (USA, 2020)
Brockmire (AMC, 2017–2020)
Creepshow (Shudder, 2019)
Daybreak (Netflix, 2019)
Devious Maids (Lifetime, 2013–2016)
Doom Patrol (HBO Max/Max, 2019–2023)
The Dukes of Hazzard (CBS, 1979–1985)
Echo (Hulu, 2024)
Falling for Angels (Here TV, 2017)
Gang Related (Fox, 2014)
Gentefied (Netflix, 2020–2021)
Get Shorty (Epix, 2017–19)
Goliath (Amazon, 2016–2021)
Good Girls (NBC, 2018–2021)
Greenleaf (OWN, 2016–2020)
Killer Women (ABC, 2014)
In Plain Sight (USA, 2008–2012)
Longmire (A&E, 2012–2014; Netflix, 2015–2017)
*Love Is_* (OWN, 2018)
Lovecraft Country (HBO, 2020)
MacGruber (Peacock, 2021)

Mayans MC (FX, 2018–2023)
Midnight, Texas (NBC, 2017–2018)
Naomi (The CW, 2022)
National Treasure: The Edge of History (Disney, 2022–2023)
Obliterated (Netflix, 2023)
Outer Range (Amazon, 2022–)
Ozark (Netflix, 2017–2022)
Perpetual Grace, Ltd. (Epix, 2019)
Preacher (AMC, 2016–2019)
Raising Dion (Netflix, 2019–2022)
Rectify (Sundance TV, 2013–2016)
Reservation Dogs (FX, 2021–2023)
Roswell, New Mexico (The CW, 2019–2022)
Rutherford Falls (Peacock, 2021–2022)
Ruthless (BET, 2020–2023)
Sistas (BET, 2019–)
Sleepy Hollow (Fox, 2013–2017)
Stranger Things (Netflix, 2016–)
Succession (HBO, 2018–2023)
Swarm (Amazon, 2023)
Teenage Bounty Hunters (Netflix, 2020)
Terminator: The Sarah Connor Chronicles (Fox, 2008–2009)
The Cleaning Lady (Fox, 2022–)
The Curse (Showtime, 2023)
The Deputy (Fox, 2020)
The Girlfriend Experience (Starz, 2016–2021)
The Lost Room (Syfy Channel, 2006)
The Messengers (CW, 2023)
The Outsider (HBO, 2020)
The Oval (BET, 2019–)
The Quad (BET, 2017–2018)
The Rookie (ABC, 2018–)
The Staircase (Netflix, 2022)
The Unsettling (Netflix, 2019)
The Walking Dead (AMC, 2010–2022)
Treme (HBO, 2010–2013)

Vida (Starz, 2018–2020)
Walker: Independence (USA, 2022–2023)
Watchmen (HBO, 2019)
Will Trent (ABC, 2023–)

NOTES

1. COLD OPENING

1. Myles McNutt, *Television's Spatial Capital: Location, Relocation, Dislocation* (New York: Routledge Press, 2022). See also Toby Miller, Nitin Govil, John McMurria, Richard Maxwell, and Ting Wang, *Global Hollywood 2*, 2nd ed. (London: British Film Institute, 2004); Michael Curtin and Kevin Sanson, eds., *Precarious Creativity: Global Media, Local Labor* (Oakland: University of California Press, 2016); and Kevin Sanson, *Mobile Hollywood: Labor and the Geography of Production* (Oakland: University of California Press, 2024). Hollywood mobility, however, has a longer history. Daniel Steinhart, in *Runaway Hollywood: Internationalizing Postwar Production and Location Shooting* (Oakland: University of California Press, 2019), argues that film location shooting began around 1910, declined with the coming of sound, and resurged after World War II for several reasons, including financial incentives. Most location shooting in the period was in the UK, Italy, and France. See also *Hollywood on Location: An Industry History*, eds. Joshua Gleich and Lawrence Webb (New Brunswick, NJ: Rutgers University Press, 2019).
2. Vicki Mayer, *Almost Hollywood, Nearly New Orleans: The Lure of the Local Film Economy* (Oakland: University of California Press, 2017), 10.
3. United States Census Bureau, "American Community Survey, 2020: ACS 5-Year Estimates Subject Table," https://data.census.gov/table/ACSST5Y2020.S1701 ?q=S1701:%20POVERTY%20STATUS%20IN%20THE%20PAST%2012%20 MONTHS&g=010XX00US$0400000,&tid=ACSST5Y2020.S1701.
4. John Gregg, "Gary Johnson says tax breaks made New Mexico the 'Second Hollywood,'" *Politifact*, July 15, 2011, www.politifact.com/factchecks/2011/jul/15/gary -johnson/gary-johnson-says-tax-breaks-made-new-mexico-secon/.
5. Aaron Couch, "New Mexico Governor Signs 'Breaking Bad' TV, Film Subsidy Bill into Law," *Hollywood Reporter*, April 4, 2013, www.hollywoodreporter.com/news /politics-news/new-mexico-governor-signs-breaking-433168/.
6. Film New Mexico, "Incentives," https://nmfilm.com/why-new-mexico/incentives -2. By contrast, California's program offers 20% incentives for new shows and 25% if a production relocates to California from elsewhere.
7. Film New Mexico, "New Mexico Tax and Revenue Payout," https://nmfilm.com /why-new-mexico/incentives-2.
8. Noelle Griffis, "The New Hollywood, 1980–1999," in *Hollywood on Location: An Industry History*, eds. Joshua Gleich and Lawrence Webb (New Brunswick, NJ: Rutgers University Press, 2019), 156.

9. John Charles Bradbury, "Film Tax Credits and the Economic Impact of the Film Industry on Georgia's Economy," Bagwell Center Policy Brief, Kennesaw State University, July 2019, https://papers.ssrn.com/sol3/papers.cfm?abstract_id=3407921.
10. Beau Evans, "Kemp Signs 2022 Georgia Budget, Adds Back Most Schol Funds Cut in Pandemic," *GPB News*, May 11, 2021, www.gpb.org/news/2021/05/11/kemp-signs-2022-georgia-budget-adds-back-most-school-funds-cut-in-pandemic. A little more than half of the cuts to public education were restored in the 2022 budget, but the rest remain in effect.
11. Gene Maddaus, "Georgia Film Credit Grows to Record $1.3 Billion," *Variety*, January 17, 2023, https://variety.com/2023/tv/news/georgia-film-credit-hits-record-1235488240/.
12. *Film LA Inc. 2019 Television Report* (Hollywood: Film L.A. Inc., 2019), 12; Todd Spangler, "Netflix is Paying Less Than $30 Million for Albuquerque Studios, Which Cost $91 Million to Build," *Variety*, October 16, 2018, https://variety.com/2018/digital/news/netflix-albuquerque-studios-deal-terms-30-million-1202981274/; and Jenny Fuster, "Atlanta Studio Sparks Protest for Plan to Clearcut 200-Acre Forest for More Soundstages," *TheWrap*, July 7, 2021, www.thewrap.com/atlanta-blackhall-studio-protests-200-acre-forest-soundstages/.
13. Mayer, *Almost Hollywood, Nearly New Orleans*, 11–12.
14. John Charles Bradbury, "Do Movie Production Incentives Generate Economic Development?," *Contemporary Economic Policy* 38, no. 2 (August 2019), 336–338.
15. Ryan Millsap, interviewed by Francesca Amiker, *11 Alive News*, Atlanta, April 27, 2018, www.11alive.com/article/news/a-look-at-the-studio-and-1-job-behind-jumanji-venom-godzilla-and-more/85-546478414.
16. Clean Tu Casa website, https://cleantucasa.com/services/tv-movie-set-cleaning/; Georgia Clean Trauma Services website, www.georgiaclean.com/services/movie-lotscene-cleanup/.
17. Blue Moon Cleaning and Restoration website, https://bluemooncleaning.com/filming-%26-entertainment.
18. Green Sweep website, https://greensweepnm.com/tv-movie-set-cleaning/.
19. Studio Cleaning Service website, www.millfac.com/film-television-production; Santa Fe's Sun Green Cleaner website, https://sungreencleaner.com/pages/about-us. Atlanta is also home to two dry cleaning businesses that serve film and TV producers: Fabricare Center, https://fcdrycleaners.com/theater-film-costume-cleaning-service/; and Cambridge Cleaners, www.cambridgecleaners2.com/.
20. Mayer, *Almost Hollywood, Nearly New Orleans*, 105.
21. See Patrick Button, "Can Tax Incentives Create a Local Film Industry? Evidence from Louisiana and New Mexico," *Journal of Urban Affairs* 43, no. 5 (2021): 658–684.
22. Cedric J. Robinson, *Black Marxism: The Making of the Black Radical Tradition* (Chapel Hill: University of North Carolina Press, 1983), 2.
23. *Histories of Racial Capitalism* (New York: Columbia University Press, 2021), edited by Destin Jenkins and Justin Leroy, does not reference *Forgeries of Memory and*

Meaning, even though Jenkins's contribution to the volume, "Debt, the New South, and the Propaganda of History," discusses *Birth of a Nation*, which is the subject of a chapter in Robinson's film book (185–211). The excellent anthology titled *Colonial Racial Capitalism*, edited by Susan Koshy, Lisa Marie Cacho, Jodi A. Byrd, and Brian Jordan Jefferson (Durham, NC: Duke University Press, 2022) does not reference *Forgeries* either, even though it includes a section on visual culture. For important exceptions to *Forgeries*' neglect in ethnic studies and Black studies, see Jordan T. Camp and Christina Heatherton, "Riots in the Master's Hall: Racism, Nationalism, and the Crisis of US Hegemony," in *Racism, Violence and Harm: Ideology, Media, and Resistance*, ed. Monish Bhatia, Scott Poynting, and Waqas Tufail (New York: Palgrave Macmillan, 2023), 225–242; and Robin D. G. Kelley, "Birth of a Nation Redux: Surveying Trumpland with Cedric Robinson," *Boston Review*, November 5, 2020, www.bostonreview.net/articles/robin-d-g-kelley-births-nation/.

24 For an exception to the neglect of Robinson in media studies, see Anamik Saha, "Production Studies of Race and the Political Economy of Media," *Journal of Cinema and Media Studies* 60, no. 1 (Fall 2020): 138–142. In "Seeing Color: The Relationship Between Popular Media and Anti-Racist Social Justice" (*The Macksey Journal* 4, no. 57 [2023]), Gabrielle Whyte draws on *Forgeries* to analyze racial stereotypes in contemporary film and television but does not engage or even mention racial capitalism.

25 Robinson, *Forgeries of Memory and Meaning: Blacks and the Regimes of Race in American Theater and Film Before World War II* (Chapel Hill: University of North Carolina Press, 2007), 180–181, 185–186.

26 Robinson, *Forgeries of Memory and Meaning*. On American Mutoscope/Biograph's host of silent anti-Indian, anti-Black, anti-Mexican, and anti-Chinese pictures, see also Curtis Marez, *University Babylon: Film and Race Politics on Campus* (Berkeley: University of California Press, 2019), 30–67.

27 Floyd C. Covington, "The Negro Invades Hollywood," *Opportunity* 7, no. 4 (April 1929), 111–113, 131, cited in Robinson, *Forgeries of Memory and Meaning*, 297–98, 380.

28 Helen Chin and Onyx Jones, "Reparations: A Journey Toward Repair," City of Culver City website, December 14, 2022, https://oag.ca.gov/system/files/media/task-force-city-culver-city-presentation-12142022-12152022.pdf; John Kent, "Culver City: From Whites Only to National Model of Diversity and Inclusion?," www.culvercity.org/files/assets/public/v/2/documents/planning-amp-development/advance-planning/speaker-series/191121_discriminatory-land-use-policies/speakerseriesdiscriminatoryslides.pdf.

29 Charlene B. Regester, "African American Extras in Hollywood During the 1920s and 1930s," *Film History* 9 (1997), 95–115.

30 Curtis Marez, *Drug Wars: The Political Economy of Narcotics* (Minneapolis: University of Minnesota Press, 2004).

31 Jodi Melamed, "Racial Capitalism," *Critical Ethnic Studies* 1, no. 1 (Spring 2015), 78.

32 Here I build on "The Prison-Televisual Complex," by Allison Page and Laurie Ouellette, *International Journal of Cultural Studies* 23, no. 1 (2019): 1–17. Focusing on a reality show filmed in prisons, they write that "the prison-televisual complex denotes the imbrication of the prison system with the television *industry*, not simply television as an ideological apparatus, or medium for *representing* prison in particular ways that uphold power dynamics. . . . The reinvention of the prison as a cultural industry . . . relies on the exploitation and commodification of racialized, poor, and dispossessed populations" (2). *Producing Precarity* similarly focuses on both industrial and representational alliances among police, prisons, and TV.

33 Herman Gray, "Recovered, Reinvented, Reimagined: *Treme*, Television Studies and Writing New Orleans," *Television & New Media* 13, no. 3 (2012), 268–278. The essay is part of a special issue on *Treme*. In the same issue, see also Helen Morgan Parmett, "Space, Place, and New Orleans on Television: From *Frank's Place* to *Treme*," 193–212; Joy V. Fuqua, "'In New Orleans, We Might Say It Like This . . .': Authenticity, Place, and HBO's *Treme*," 235–242; L. L. Thomas, "'People Want to See What Happened': *Treme*, Televisual Tourism, and the Racial Remapping of Post-Katrina New Orleans," 213–224; and W. Rathke, "*Treme* for Tourists: The Music of the City without the Power," 261–267.

34 Helen Morgan Parmett, *Down in Treme: Race, Place, and New Orleans on Television* (Stuttgart: Franz Steiner Verlag Wiesbaden GmbH, 2019), 16–17. See also her essay, "Site-Specific Television as Urban Renewal: Or, How Portland became Portlandia," *International Journal of Cultural Studies* 21, no. 1 (2018): 42–56.

35 Mayer, *Almost Hollywood, Nearly New Orleans*, 79.

36 Gilmore, for example, analyzes how state and capitalist investments in the San Joaquin Valley of California have (re)produced racist places where people of color are vulnerable to premature death, while Dorr analyzes how music corporations and grass-roots musical performers in Latin American produce antagonistic places. Finally, Sánchez and Pita present a critical history of Chicanx literature's representation of processes of enclosure in the US southwest. See Ruth Wilson Gilmore, *Golden Gulag: Prisons, Surplus, Crisis, and Opposition to Globalization* (Berkeley: University of California Press, 2007); Kirstie Dorr, *On Site, In Sound: Performance Geographies in Latina/o America* (Durham, NC: Duke University Press, 2018); and Rosaura Sánchez and Beatrice Pita, *Spatial and Discursive Violence in the U.S. Southwest* (Durham, NC: Duke University Press, 2021).

37 Laura Pulido, Laura Barraclough, and Wendy Cheng, *A People's Guide to Los Angeles* (Berkeley: University of California Press, 2012), 7–8.

38 Ibid., 7.

39 Ibid.

40 Kevin J. Delaney and Rick Eckstein, *Public Dollars, Private Stadiums: The Battle over Building Sports Stadiums* (New Brunswick, NJ: Rutgers University Press, 2003), Kindle Location 2775–6. See also Michael N. Danielson, *Home Team: Professional Sports and the American Metropolis* (Princeton, NJ: Princeton University Press, 1997), and James T. Bennett, *They Play, You Pay: Why Taxpayers Build Ball-*

parks, Stadiums, and Arenas for Billionaire Owners and Millionaire Players (New York: Springer, 2012).

41 George Lipsitz, *How Racism Takes Place* (Philadelphia: Temple University Press, 2011), 73–74.

42 On the limits of textual reductionism in media studies, see Toby Miller, Nitin Govil, John McMurria, and Richard Maxwell, *Global Hollywood* (London: BFI Publishing, 2001), 14–15.

43 John Thornton Caldwell, *Production Culture: Industrial Reflexivity and Critical Practice in Film and Television* (Durham, NC: Duke University Press, 2008); Vicki Mayer, Miranda J. Banks, and John T. Caldwell, eds., *Production Studies: Cultural Studies of Media Industries* (New York: Routledge, 2009); Vicki Mayer, *Below the Line: Producers and Production Studies in the New Television Economy* (Durham, NC: Duke University Press, 2011); Miranda Banks, Bridget Conor, and Vicki Mayer, eds., *Production Studies, The Sequel!: Cultural Studies of Global Media Industries* (New York: Routledge, 2015); Michael Curtin and Kevin Sanson, eds., *Precarious Creativity: Global Media, Local Labor* (Oakland: University of California Press, 2016); Myles McNutt, *Television's Spatial Capital: Location, Relocation, Dislocation* (New York: Routledge Press, 2022); Kevin Sanson, *Mobile Hollywood: Labor and the Geography of Production* (Oakland: University of California Press, 2024).

While most production studies focus on media workers and not the kinds of ancillary labor that TV production also requires, there are some important exceptions. See, for example, Mari Paredes Castañeda, "Television Set Production at the US–Mexico Border: Trade Policy and Advanced Electronics for the Global Market," in *Critical Cultural Policy: A Reader*, ed. Justin Lewis and Toby Miller (Malden, MA: Blackwell, 2002), 272–281; and Lisa Parks, "Falling Apart: Electronics Salvaging and the Global Media Economy," in *Residual Media*, ed. Charles Acland (Minneapolis: University of Minnesota Press, 2007), 32–47.

44 Steven D. Classen, *Watching Jim Crow: The Struggles over Mississippi TV, 1955–1969* (Durham, NC: Duke University Press, 2004); Faye Ginsburg, Brian Larkin, and Lila Abu-Lughod, *Media Worlds: Anthropology on New Terrain* (Berkeley: University of California, 2002); Herman Gray, *Watching Race: Television and the Struggle for Blackness* (Minneapolis: University of Minnesota Press, 2004) and *Cultural Moves: African Americans and the Politics of Representation* (Berkeley: University of California Press, 2005); Devorah Heitner, *Black Power TV* (Durham, NC: Duke University Press, 2013); Chon Noriega, *Shot in America: Television, the State, and the Rise of Chicano Cinema* (Minneapolis: University of Minnesota Press, 2000); Yeidy Rivero, *Tuning Out Blackness: Race and Nation in the History of Puerto Rican Television* (Durham, NC: Duke University Press, 2005); Dustin Tahmahkera, *Tribal Television: Viewing Native People in Sitcoms* (Chapel Hill: University of North Carolina Press, 2014); and Sasha Torres, *Black, White, and in Color: Television and Black Civil Rights* (Princeton, NJ: Princeton University Press, 2003).

45 Jodi Melamed, "Diversity," in *Keywords for American Cultural Studies*, eds. Bruce Burgett and Glenn Hendler (New York: New York University Press, 2020), 93–97. See also Nick Mitchell, "Diversity," in *Keywords for African American Studies*, edited by Erica R. Edwards, Roderick A. Ferguson, and Jeffrey O. G. Ogbar (New York: New York University Press, 2018), 69–74.

46 Quoted by Reece Ristau, "Eva Longoria Talks 'Devious Maids' Backlash, Latino Perceptions at Produced By Conference," *Variety*, May 31, 2015, https://variety.com/2015/tv/news/eva-longoria-devious-maids-latino-1201509025/.

47 Quoted by Jeanne Jakle, "Longoria Lauds 'Devious Maids' Amid Heated Debate," *My San Antonio*, June 25, 2013, www.mysanantonio.com/entertainment/entertainment_columnists/jeanne_jakle/article/Longoria-lauds-Devious-Maids-amid-heated-debate-4613193.php.

48 For data on the underrepresentation of BIPOC creatives, see Ana-Christina Ramón, Michael Tran, and Darnell Hunt, *Hollywood Diversity Report 2023, Part 2: Television* (Los Angeles: UCLA Entertainment and Media Research Initiative, 2023).

49 Bambi Haggins, *Laughing Mad: The Black Comic Persona in Post-Soul America* (New Brunswick, NJ: Rutgers University Press, 2007). For critical accounts of the industrial limits and burden of representation faced by Black and Latinx creatives, see Raquel J. Gates, *Double Negative: The Black Image and Popular Culture* (Durham, NJ: Duke University Press, 2018); and Isabel Molina-Guzmán, *Latinas and Latinas on TV: Colorblind Comedy in the Post-racial Network Era* (Tucson: University of Arizona Press, 2018).

50 *Hollywood Diversity Report 2023*, 16, 20–21.

51 Felicia D. Henderson, "From Sitcom Girl to Drama Queen: Soul Food's Showrunner Examines Her Role in Creating TV's First Successful Black-Themed Drama," in *Watching While Black Rebooted!: The Television and Digitality of Black Audiences*, ed. Beretta E. Smith-Shomade (New Brunswick, NJ: Rutgers University Press, 2023), 57–55. See also Ralina L. Joseph, *Postracial Resistance: Black Women, Media, and the Uses of Strategic Ambiguity* (New York: New York University Press, 2018), 83–107; and Robin R. Means Coleman and Andre M. Cavalcante, "Two Different Worlds: Television as a Producer's Medium," in *Watching While Black: Centering the Television of Black Audiences*, ed. Smith-Shomade (New Brunswick, NJ: Rutgers University Press, 2012).

52 Means and Cavalcante, 46–7.

53 Ibid., 66.

54 I am grateful to Julia Lesage and Michael Litwack for suggesting this connection.

55 Robinson, *Forgeries of Memory and Meaning*, 227.

56 Ibid., 231.

57 Ibid., 271.

58 Ibid., 231.

59 Closest to my approach here is Pedro A. Regalado's "'They Speak Our Language . . . Business': Latinx Businesspeople and the Pursuit of Wealth in New

York," in *Histories of Racial Capitalism*. He argues that, although scholars have assumed racial capitalism "has largely been a story of white people," Robinson's concept can be used to analyze the history of how Latinx elites leveraged government subsidies to accumulate wealth in the name of lifting Latinx people out of poverty via capitalist development (231–250). Similarly, in their introduction to *Colonial Racial Capitalism*, Koshy, Cacho, Byrd, and Jefferson suggest analyzing post-colonial elites in terms of racial capitalism (4).

60 I don't study reality TV, a topic which could be the basis for its own book.
61 Clyde Woods, "Les Misérables of New Orleans: Trap Economics and the Asset Stripping Blues, Part 1," *American Quarterly* 61, no. 3 (September 2009), 769–796.
62 Quoted in Lea Palmiera, "'Get Shorty' Season 3," *Decider*, October 4, 2019, https://decider.com/2019/10/04/chris-odowd-interview-get-shorty-epix/.

2. POOR LOCATIONS

1 Dara Orenstein, *Out of Stock* (Chicago: University of Chicago Press, 2019), 13–14.
2 Ellie Hensley, "Why Atlanta's warehouse owners are jumping into the movie business," *The Business Journals*, November 1, 2014, www.bizjournals.com/bizjournals/news/2014/11/01/why-atlantas-warehouse-owners-are-jumping-into-the.html.
3 Jason Strykowski, *A Guide to New Mexico Film Locations: From Billy the Kid to Breaking Bad and Beyond* (Albuquerque: University of New Mexico Press, 2021), 162.
4 Chris Wilson, "The Historic Railroad Buildings of Albuquerque: An Assessment of Significance," Wheels Museum, Albuquerque, NM, 1.
5 Third Rail Studios, "Third Rail Studios Info Sheet," https://thirdrailstudios.com/wp-content/uploads/Third-Rail-Studios-Info-Sheet-20190901.pdf.
6 Adam Bruns, "Electric Owl Studios Comes In For a Landing," *Site Selection Magazine*, March 2022, https://siteselection.com/investor-watch/electric-owl-studios-comes-in-for-a-landing.cfm.
7 Electric Owl Studios website, https://electricowlstudios.com.
8 Ipek A. Celik Rappas, "From *Titanic* to *Game of Thrones*: Promoting Belfast as a Global Media Capital," *Media, Culture, and Society* 41, no. 1 (January 2019): 3.
9 Diane Wei Lewis, "'The Longed-For Crystal Palace," in *In the Studio: Visual Creation and its Material Environments*, ed. Brian R. Jacobson (Oakland: University of California Press, 2020), 31.
10 Rielle Navitski, "Regulating Light, Interiors, and the National Image: Electrification and the Studio Space in Silent-era Brazil," in Jacobson, *In the Studio*, 43.
11 Lynn Spigel, "Made-for-Broadcast Cities," in Jacobson, *In the Studio*, 215.
12 "Third Rail Studios Info Sheet."
13 "EUE/Screen Gems Studios Atlanta" (booklet), https://euescreengems.com/wp-content/uploads/2019/01/EUE-Brochure_Book_PDF_rev12_28_18.pdf.
14 Shadowbox Studios website, https://shadowboxstudios.com.
15 Dan Immergluck, *Red Hot City: Housing, Race, and Exclusion in Twenty-First Century Atlanta* (Oakland: University of California Press, 2022), 50-52.

16 Santa Fe Studios website, www.santafestudios.com/our-facilities/offices/index.html.
17 Mike Fleming Jr., "Netflix Commits $1 Billion More In New Mexico Production Funding As It Expands ABQ Studios; 'Stranger Things' Joins List Of Albuquerque-Set Shows," *Deadline*, November 23, 2020, https://deadline.com/2020/11/netflix-billion-dollar-production-commitment-new-mexico-abq-studios-stranger-things-1234620435/.
18 City of Albuquerque, "City Making Major Upgrades, Improving Access to Mesa Del Sol," 2020, www.cabq.gov/municipaldevelopment/news/city-making-major-upgrades-improving-access-to-mesa-del-sol.
19 Bruce Krasnow and Dan Schwartz, "Santa Fe Studios Clashes With County Over Debt Repayment, Expansion," *Santa Fe New Mexican*, December 13, 2015, www.santafenewmexican.com/news/local_news/santa-fe-studios-clashes-with-county-over-debt-repayment-expansion/article_f6e830fa-a308-5e6b-8285-d3d4fdb18a41.html.
20 Maria Saporta, "Fort Mac-Tyler Perry Deal Could Close Friday; Sen. Vincent Fort Declares Sale Illegal," *Atlanta Business Chronicle*, June 26, 2015, www.bizjournals.com/atlanta/morning_call/2015/06/fort-mac-tyler-perry-deal-could-close-friday-sen.html.
21 Tyler Estep, "Officials: 'Historic' Film Studio Expansion to Bring 2,400 Jobs to South DeKalb, Development Authority Grants $68 Million Incentive Package," *Atlanta Journal-Constitution*, April 14, 2022, www.ajc.com/neighborhoods/dekalb/officials-historic-film-studio-expansion-to-bring-2400-jobs-to-south-dekalb/W6BZIEI2SFEC5IIGMRSN45KKJY/.
22 Fuster, "Atlanta Studio Sparks Protests."
23 Ibid.
24 Alyse Eady, "Georgia's Tire Problem," *Good Day Atlanta*, Fox 5, June 29, 2018, www.youtube.com/watch?v=65krwSKXdRE.
25 "Park Closure Sparks Talk of Ongoing Lawsuit Against DeKalb and Blackhall," *The Champion*, July 25, 2020, https://thechampionnewspaper.com/park-closure-sparks-talk-of-ongoing-lawsuit-against-dekalb-and-blackhall/.
26 Rob DiRienzo, "Anonymous Activist Claims Responsibility for Blaze at DeKalb Film Studio: 'May This be a Warning to Them,'" *Fox 5 Atlanta*, October 21, 2022, www.fox5atlanta.com/news/anonymous-activist-claims-responsibility-for-blaze-at-dekalb-film-studio-may-this-be-a-warning-to-them.
27 Lucia Gravel, "Forest Defenders Continue Fight Against 'Cop City,'" *The Southerner*, Midtown High School, Atlanta, October 19, 2022, https://thesoutherneronline.com/90048/news/forest-defenders-continue-fight-against-cop-city/#modal-photo. \; https://www.wsbtv.com/news/local/dekalb-county/business-owners-want-dekalb-leaders-do-something-about-crimes-committed-against-them/REHLLCYK2VCJNOXGQUA73WWQVI/.
28 Robin Wood, "An Introduction to American Horror," in *Robin Wood on the Horror Film: Collected Essays and Reviews*, ed. Barry Keith Grant, 114–67 (Detroit: Wayne State University Press, 2018), 150.

29 Like RCA's electronics manufacturing—including of televisions—film and TV producers profit from moving to poor regions with high unemployment and poor women workers. See Jefferson Cowie, *Capital Moves: RCA's Seventy-Year Quest for Cheap Labor* (Ithaca, NY: Cornell University Press, 2001).
30 Albuquerque Development Commission, "LEDA-20-7: Netflix Studio Project," report for Local Economic Development Act Hearing, November 23, 2020, www.cabq.gov/mra/documents/case-2020-10-leda-20-7-staff-analysis-netflix-studios.pdf.
31 Patrick Sisson, "Albuquerque is Winning the Streaming Wars," *Bloomberg*, May 3, 2021, www.bloomberg.com/news/articles/2021-05-03/why-hollywood-is-moving-to-albuquerque.
32 Georgia Film Office, "Devious Maids: Three Seasons in Atlanta," 2016, www.youtube.com/watch?v=BDTcX7vXjgg&t=3s.
33 Georgia Film Office, "Devious Maids Stars on Southern Hospitality, and Food," www.youtube.com/watch?v=syEmMof8Qbc&t=120s.
34 Georgia Film Office, "Devious Maids Loves the ATL! (It's Mutual)," 2015, www.youtube.com/watch?v=rt6L_yYWoJ8.
35 Jonathan Banks, Bryan Cranston, Vince Gilligan, Gennifer Hutchinson, and Michael Slovis, "DVD Commentary on 'No Mas,'" *Breaking Bad*, S3E1, AMC, 2010.
36 I am grateful to an anonymous NYU Press reader for this formulation.
37 Brian R. Jacobson, "Introduction: Studio Perspectives," in *In the Studio*, 7.
38 Jennifer Faye, *Inhospitable World: Cinema in the Time of the Anthropocene* (New York: Oxford University Press, 2018), 16.
39 Ibid., 9.
40 Mayer, *Almost Hollywood, Nearly New Orleans*, 46.
41 Gloria Tatum, "Native Americans Share Concerns Over Fate of Forest," *Streets of Atlanta*, May 2, 2022, https://streetsofatlanta.blog/2022/05/02/native-americans-share-concerns-over-fate-of-forest/; Sukainah Abid-Kons, "Fight Against 'Cop City' Continues at the 'Community in Weelaunee' Summit," *Georgia Voice*, May 5, 2022, https://thegavoice.com/community/fight-against-cop-city-continues-at-the-community-in-weelaunee-summit/; Stephanie Wakefield and Glenn Dyer, "Stop the Metaverse, Save the Real World," *e-flux*, September, 2022, www.e-flux.com/architecture/horizons/493130/stop-the-metaverse-save-the-real-world/; Mark Auslander and Avis E. Williams, "What these Trees have Seen: Slavery, Post-Slavery, and Anti-Blackness in the South River (Weelaunee) Forest Zone," *Mark Auslander: New Visons on Museums, Community Engagement, Art, and Science in the Public Interest* (website), April 23, 2022, https://markauslander.com/2022/08/04/what-these-trees-have-seen-slavery-post-slavery-and-anti-blackness-in-the-south-river-welaunee-forest-zone/; and Ella Fassler, "Activists Occupying The Woods to Block 'Cop City' Face Terrorism Charges," *Vice*, December 21, 2022, www.vice.com/amp/en/article/xgy9yk/activists-occupying-the-woods-to-block-cop-city-face-terrorism-charges.
42 Immergluck, *Red Hot City*, 42.

43 Curtis Marez, "'Plenty of Room to Swing a Rope': *Watchmen* and the Racial Politics of Place," in *After Midnight: Watchmen After Watchmen*, ed. Drew Morton (Jackson: University Press of Mississippi, 2022), 146–7.
44 New Mexico Indian Affairs Department, "History: New Mexico's Twenty-Three Tribes and the Indian Affairs Department," www.iad.state.nm.us/about-us/history/.
45 Myles McNutt, *Television's Spatial Capital: Location, Relocation, Dislocation* (New York: Routledge, 2022), 264–5.
46 Quoted by Mayers, *Almost Hollywood, Nearly New Orleans*, 67.
47 Noelle Griffis, "The New Hollywood, 1980–1999," in *Hollywood on Location: An Industry History*, eds. Joshua Gleich and Lawrence Webb (New Brunswick, NJ: Rutgers University Press, 2019), 158.
48 Mickey O'Connell, "'Reservation Dogs' Boss on Combatting Indigenous Stereotypes, Embracing Gripes and That Emmy Snub," *Hollywood Reporter*, August 18, 2022, www.hollywoodreporter.com/tv/tv-features/reservation-dogs-sterlin-harjo-emmy-snub-indigenous-stereotypes-1235199618/.
49 Kate Aurthur, "From Daniel's Death to Native Humor, 'Reservation Dogs' Showrunner Sterlin Harjo Answers Our Burning Questions," *Variety*, June 15, 2022, https://variety.com/2022/tv/news/reservation-dogs-sterlin-harjo-daniel-death-native-humor-1235294333/.

3. CRIME TV

1 Albuquerque Police Department, Facebook, December 13, 2016, www.facebook.com/abqpolice/photos/a.482215295149670/1216874778350381/?type=3.
2 Claire McNear, "Ranking the Cameos From 'Breaking Bad' and 'Better Call Saul' in 'El Camino,'" *The Ringer*, October 15, 2019, www.theringer.com/movies/2019/10/15/20914631/el-camino-netflix-cameos-ranked-breaking-bad-better-call-saul; Bill Bradley, "What Happened to Skinny Pete in 'El Camino'? This Real-Life Police Officer Knows," *Huffpost*, October 17, 2019, www.huffpost.com/entry/what-happened-skinny-pete-el-camino-police_n_5da109f4e4b06ddfc519f0e1.
3 Page and Ouellette, "The Prison-Televisual Complex," 3.
4 Elise Kaplan, "APD: Former Spokesman Simon Drobik was 'Gaming the System,'" *Albuquerque Journal*, October 23, 2020, www.abqjournal.com/news/local/apd-former-spokesman-simon-drobik-was-gaming-the-system/article_76a55a69-4937-5ea2-8223-1e6d7f5100e1.html.
5 John Honeycutt, "Albuquerque Reaches $42.5K Settlement in Excessive Force Case," *KRQE News*, September 1, 2023, www.krqe.com/news/albuquerque-metro/albuquerque-reaches-42-5k-settlement-in-excessive-force-case/.
6 Chacon v. Albuquerque Police Department, 347 F. Supp. 3d 694 (D.N.M. 2018), August 31, 2018, *Casetext*, https://casetext.com/case/chacon-v-albuquerque-police-dept.
7 Elsie Kaplan, "APD Officer Violated Policies in Inmate's Suicide," *Albuquerque Journal*, October 6, 2020, www.abqjournal.com/news/local/apd-officer-violated-policies-in-inmates-suicide/article_73d18188-e036-5597-854f-238003a68089.html.

8 Nick Pinto, "When Cops Break Bad: Inside a Police Force Gone Wild," *Rolling Stone*, January 29, 2015, 5–6, www.rollingstone.com/culture/culture-news/when-cops-break-bad-inside-a-police-force-gone-wild-69487/.

9 Nadia Whitehead, "What Was it Like to Consult for Breaking Bad?," *Science*, May 2, 2014, www.science.org/content/article/what-was-it-consult-breaking-bad.

10 Jillian Wooten, "An Historical Analysis of The Atlanta Prison Farm," Atlanta City Planning Office, November 5, 1999, https://dekalbhistory.org/wp-content/uploads/2019/11/historical-analysis-of-honor-farm.pdf; www.ajc.com/blog/buzz/old-atlanta-prison-farm-favorite-with-filmmakers/yHzdSgh4ywGlzFaxjokqdL/.

11 Jacqueline Echols, interviewed by Sorrel Inman, "Environmental Justice in the South River Watershed," *Mergoat Magazine*, Spring 2023, 54–55, https://issuu.com/mergoat/docs/atl_section_-_blue_hollers_-_digital/48.

12 Christina Maxouris, "Atlanta Wants to Build a Massive Police Training Facility in a Forest. Neighbors are Fighting to Stop It," *CNN*, September 24, 2022, www.cnn.com/2022/09/24/us/atlanta-public-safety-training-center-plans-community/index.html.

13 Jennifer Brett, "Old Atlanta Prison Farm a Favorite with Filmmakers," *Atlanta Journal-Constitution*, October 31, 2014, www.ajc.com/blog/buzz/old-atlanta-prison-farm-favorite-with-filmmakers/yHzdSgh4ywGlzFaxjokqdL/.

14 Film New Mexico, "Production Resources: State Owned Buildings," https://nmfilm.com/filmmaker-resources/permits-procedures/state-owned-buildings.

15 New Mexico Corrections Department, "Old Main: Filming and Tours," www.cd.nm.gov/old-main/.

16 Gwinnett Film, Location Search Results for Prisons/Jails, December 8, 2024, https://gwinnett.locationshub.com/search_results.aspx?search=&search_mode=and&ctry=US&state=GA&city=&bor=&cat=989&subcat=989&style=&lname=&lid=&int=&sort=date&view=16.

17 "Local Jail Reaps Benefits of Georgia's Booming Film Industry," *WSB-TV*, November 26, 2016, www.youtube.com/watch?v=oAi9MpbwDpU.

18 Chief J. E. Edens, Senoia Police Department Memorandum, July 8, 2019, www.senoia.com/sites/default/files/fileattachments/city_council/meeting/4411/vehicle_purchase_request.pdf.

19 Jessica Hatrick and Olivia González "*Watchmen*, Copaganda, and Abolition Futurities in US Television," *Lateral: Journal of The Cultural Studies Association* 11, no. 2 (Fall 2022), https://csalateral.org/issue/11-2/watchmen-copaganda-abolition-futurities-us-television-hatrick-gonzalez/. See also Andrew Hoberek, "Of Watchmen and Great Men: The Graphic Novel, The Television Series, and the Police," *Literature/Film Quarterly* 49, no. 4 (Fall 2021), https://lfq.salisbury.edu/_issues/49_4/of_watchment_and_great_men_the_graphic_novel_the_television_series_and_the_police.html.

20 Color of Change and the USC Annenberg Norman Lear Center, "Normalizing Injustice," January 2020, https://colorofchange.org/press_release/normalizing-injustice-new-landmark-study-by-color-of-change-reveals-how-crime-tv-shows-distort-understanding-of-race-and-the-criminal-justice-system/.

21 In both states, Latinx men and women and Black men and women are overrepresented in prisons, while white men and women are underrepresented. See Derek Chin, "Profile of New Mexico Prison Population," New Mexico Sentencing Commission, December 2021, https://nmsc.unm.edu/reports/2021/profile-of-new-mexico-prison-population.pdf; and "Georgia Profile," Prison Policy Initiative, www.prisonpolicy.org/profiles/GA.html.
22 Donovan J. Thomas, "Families of Three Inmates Who Died at Gwinnett Jail Push for Changes," *Atlanta Journal-Constitution*, October 15, 2022; Barron Jones and Lalita Moskowitz, "The Inhumane Conditions at MDC are a Result of Over-Incarceration," *ACLU New Mexico*, July 15, 2020, www.ajc.com/news/families-of-3-inmates-who-died-at-gwinnett-jail-push-for-changes/MNHK6UJXD5BOZG4H5PGH65NYZQ/.
23 Life here imitated art, when on April 24, 2020, months after *Deputy*'s release, the actual Los Angeles Sheriff Alex Villanueva banned the practice of transferring local immigrant inmates to federal immigration authorities without a warrant.
24 Ariana Romero, "The True Story of *Teenage Bounty Hunters*, Straight From Its Creator," *Refinery 29*, August 14, 2020, https://ew.com/tv/teenage-bounty-hunters-netflix-kathleen-jordan-preview/; https://www.refinery29.com/en-us/2020/08/9966810/is-teenage-bounty-hunters-true-story-netflix.
25 Marez, *Drug Wars*, 27–29.
26 Quoted by Anthony D'Alessandro, "Emmys: 'Devious Maids' Second Life After its Death at ABC; Creator Marc Cherry Q&A," *Deadline*, June 7, 2014, https://deadline.com/2014/06/emmys-devious-maids-second-life-after-its-death-at-abc-creator-marc-cherry-qa-741969/.
27 Jillian Báez, "Television for All Women? Watching Lifetime's *Devious Maids*," in *Cupcakes, Pinterest, and Lady Porn: Feminized Popular Culture in the Early Twenty-First Century*, ed. Elana Levine (Chicago: University of Illinois Press, 2015), 56.
28 Max Gao, "A TV Drama in ICE Custody," *Los Angeles Times*, February 8, 2022, E1, E6.
29 Diep Tran, "'The Cleaning Lady' Aims to Break Stereotypes About Asian Workers in the US," *NBC News*, January 31, 2022, www.nbcnews.com/news/asian-america/-cleaning-lady-aims-break-stereotypes-asian-workers-us-rcna14270#:~:text ="There%20is%20actually%20a%20greater,to%20center%20a%20Filipino%20family.

4. HORROR TV

1 Sharareh Drury, "'Walking Dead' Showrunner on How the Pandemic Impacts Its Final Season," *Hollywood Reporter*, August 13, 2021, www.hollywoodreporter.com/tv/tv-news/walking-dead-final-season-preview-angela-kang-1234995522/.
2 Jeremy A. Golden, Karen K. Wong, Christine M. Szablewski, et al., "Characteristics and Clinical Outcomes of Adult Patients with Covid-19—Georgia March 2020," Centers for Disease Control and Prevention, April 29, 2020, www.cdc.gov/mmwr/volumes/69/wr/mm6918e1.htm; Jazmyn T. Moore, Jessica N. Ricaldi, Charles E. Rose, et al. "Disparities in Incidence of Covid-19 Among Under-

represented Racial/Ethnic Groups in Counties Identified as Hotspots During June 5–18, 2020–22 States, February-June 2020," Centers for Disease Control and Prevention, August 12, 2020, www.cdc.gov/mmwr/volumes/69/wr/mm6933e1.htm#:~:text=During%20June%205–18%2C%20205,more%20underrepresented%20racial%2Fethnic%20groups.

3 Jerry Shannon, Amanda Abraham, Grace Bagwell Adams, and Mathew Hauer, "Racial Disparities for Covid 19 Mortality in Georgia: Spatial Analysis. By Age Based on Excess Deaths," *Social Science and Medicine*, January 2022, www.ncbi.nlm.nih.gov/pmc/articles/PMC8734109/#:~:text=Conclusions,to%20social%20determinants%20of%20health.

4 On the consequences of the COVID-19 pandemic for film and TV production, see Kate Fortmueller, *Hollywood Shutdown: Production, Distribution, and Exhibition in the Time of COVID* (Austin: University of Texas Press, 2021), especially 26–31, where she discusses Georgia's early resumption of production.

5 Clyde Woods, "Les Misérables of New Orleans: Trap Economics and the Asset Stripping Blues, Part 1," *American Quarterly* 61, no. 3 (September 2009), 769–796.

6 Stephanie A. Schwartz, "Consciousness, Covid, and the Rise of an American Death Cult," *Explore* 18, no. 3 (May–June 2022), 259–263, www.ncbi.nlm.nih.gov/pmc/articles/PMC8935966/; Jennifer Rubin, "How the Republican Party Became a Death Cult," *The Washington Post*, June 23, 2020, www.washingtonpost.com/opinions/2020/06/23/how-republican-party-became-death-cult/; and Jonathan Chait, "American Death Cult: Why has the Republican Response to the Pandemic Been so Mind-Bogglingly Disastrous?," *New York Magazine*, July 20, 2020, https://nymag.com/intelligencer/2020/07/republican-response-coronavirus.html.

7 Daniel Victor, "Hart Family Parents Killed 6 Children in Murder-Suicide, Jury Decides," *New York Times*, April 5, 2019, www.nytimes.com/2019/04/05/us/hart-family-murder-suicide.html.

8 Kelsie Barton, *The Atlantan*, March 29, 2021, https://atlantanmagazine.com/exclusive-peek-at-future-plans-atl-bouckaert-farm.

9 Madeline Leung Coleman, "Donald Glover Hoped *Swarm* Would Make You Uncomfortable," *Vulture*, March 28, 2023, www.vulture.com/article/donald-glover-swarm-interview-dre-dominique-fishback.html.

10 Ibid.

11 Bertrand Cooper, "Who Actually Gets to Create Black Pop Culture?," *Current Affairs*, July 25, 2021, www.currentaffairs.org/2021/07/who-actually-gets-to-create-black-pop-culture.

12 Hillary Atkin, "Swarm Co-Creator Janine Nabers on Working With Donald Glover, Billie Eilish, and Fellow Emmy Nominee Dominique Fishback," *Above the Line*, August 14, 2022, https://abovetheline.com/2023/08/14/swarm-janine-nabers-interview-donald-glover-billie-eilish-dominique-fishback/.

13 On the limits of colorblind casting, see Kristen J. Warner, *The Cultural Politics of Colorblind TV Casting* (New York: Routledge, 2015).

14 Lynne Joyrich argues that *Sleepy Hollow* is contradictory, criticizing "the naturalization of the ideology of American exceptionalism even as it appears to endorse it." See "American Dreams and Demons: Television's 'Hollow' Histories and Fantasies of Race," *The Black Scholar* 48, no. 1 (2018), 34.
15 Maureen Ryan, *Burn It Down: Power, Complicity, and a Call for Change in Hollywood* (New York: Mariner Books, 2003), 129–161.
16 Denise Petski, "Nichole Beharie Speaks Out About Her 'Sleepy Hollow' Exit and the Controversy's Impact on her Career," *Deadline*, June 20, 2020, https://deadline.com/2020/06/nicole-beharie-sleepy-hollow-exit-controversy-impact-career-1202965074/; Sonaiya Kelley, "Nichole Beharie was Called 'Problematic' and Blacklisted. 'Miss Juneteenth' Brings Redemption," *San Diego Union-Tribune*, June 19, 2020, www.sandiegouniontribune.com/entertainment/movies/story/2020-06-19/nicole-beharie-miss-juneteenth.
17 Stacy Lambe, "How 'Antebellum,' 'Lovecraft County' and More Horror Stories are Centering Black Experiences," *Entertainment Tonight*, September 25, 2020, www.etonline.com/how-antebellum-lovecraft-country-and-more-horror-stories-are-centering-black-experiences-153705.
18 On Hippolyta's journey, see Jordan Harper and Henry Jenkins, "Confronting Horror, Embracing Fantasy: A Conversation about Lovecraft Country and Radical Imagination in Higher Education," *Policy Futures in Education* 20, no. 1 (2022), 76, 78; and DeLisa D. Hawkes, "Hippolyta's Spiritual Awakening Through Spiritual Warfare in *Lovecraft Country* (2020)," *First Thoughts on Lovecraft Country*, a special issue of *Studies in the Fantastic* 12 (Winter 2021): 1–17.
19 Stacy Lambe, "'Lovecraft Country' Creator Says There are 'Seasons Upon Seasons' of Stories Left to Tell," *Entertainment Tonight*, October 18, 2020, www.etonline.com/lovecraft-country-creator-on-future-seasons-and-changes-from-book-154873.
20 Ibid.
21 Joseph Lewis, "Man's Fear of a Black Planet: Monstrous Ontological Encounters at Sundown in HBO's *Lovecraft Country*," *First Thoughts on Lovecraft Country*, a special issue of *Studies in the Fantastic* 12 (Winter 2021): 26.
22 Robin Wood, "An Introduction to American Horror," 114–67.

5. SCIENCE FICTION TV

1 Bree Newsom, "Go Ahead, Topple the Monuments to the Confederacy, All of Them," *Washington Post*, August 18, 2017, www.washingtonpost.com/opinions/go-ahead-topple-the-monuments-to-the-confederacy-all-of-them/2017/08/18/6b54c658-8427-11e7-ab27-1a21a8e006ab_story.html.
2 For an account of its popularity and critical acclaim, see Benjamin Burroughs, Benjamin J. Morse, Travis Snow, and Michael Carmona, "The Masks We Wear: *Watchmen*, Infrastructural Racism, and Anonymity," *Television and New Media* 23, no. 3 (2023): 247–263.
3 Meghan O'Keefe, "HBO's 'Watchmen' was Ahead of its Time," *Decider*, June 4, 2020, https://decider.com/2020/06/04/watchmen-on-hbo-2020-relevance-tulsa

-massacres/; Ray Flook, "Watchmen: Trump Makes Life Feel More Like Deleted Scenes Than Season 2," *Bleeding Cool*, January 7, 2021. https://bleedingcool.com/tv/watchmen-trump-makes-life-feel-more-like-deleted-scenes-than-season-2/; Astead W. Herndon, "Black Tulsans, With a Defiant Juneteenth Celebration, Send a Message to Trump," *New York Times*, June 19, 2020, https://www.nytimes.com/2020/06/19/us/politics/juneteenth-tulsa-trump-rally.html.

4 Wanzo, in Gillespie et al., "Thinking about *Watchmen*." See also Nicole Simek, "Speculative Futures: Race in Watchmen's Worlds," *Symploke* 28, no. 1–2 (2020): 385–404; "*Watchmen*: From Co-Mix to Remix," a special issue of *Literature/Film Quarterly* 49, no. 4 (Fall 2021); Drew Morton, ed., *After Midnight: Watchmen After Watchmen* (Jackson: University Press of Mississippi, 2022).

5 According to the 2000 US Census, Cedartown's population is 20.20% Black and 22.4% Latinx.

6 West Georgia Textile Heritage Trail, "Cedartown," https://westgatextiletrail.com/cedartown/.

7 Myrick, "No Plans."

8 Kirk Savage, *Standing Soldiers, Kneeling Slaves: Race, War, and Monument in Nineteenth-Century America* (Princeton, NJ: Princeton University Press, 1999).

9 Del Upton, *What Can and Can't be Said: Race, Uplift, and Monument Building in the Contemporary South* (New Haven, CT: Yale University Press, 2015), Kindle Location 598.

10 Ibid., Kindle Location 493. See also Savage, *Standing Soldiers, Kneeling Slaves*. On confederate monuments in Georgia, see David N. Wiggins, *Georgia's Confederate Monuments and Cemeteries* (New York: Arcadia Publishing, 2006). For Confederate graves in the Cedartown area, see "Polk County Georgia Archives Cemeteries," USGenWeb Archives, https://thegagenweb.com/gagwinnett/album/confedmem.html. Finally, I am indebted to *All Monuments Must Fall: A Collaboratively Produced Syllabus*, https://allmonumentsmustfall.com.

11 Historic Oakland Foundation, "Character Areas and Landmarks."

12 Friends of Decatur Cemetery, "Lives That Made Our City: Decatur Cemetery Walking Tour," Decatur, GA, n.d., www.decaturga.com/media/21356.

13 Faye Campbell, "Multifaceted Meanings of Monuments Continue to Evolve," *Newnan Times-Herald*, June 20, 2020, www.times-herald.com/news/local/multi-faceted-meanings-of-monuments-continue-to-evolve/article_da29e9eb-ae9e-51d8-8f1d-48b0ae6f33f0.html.

14 Chris Joyner, "Georgia Capitol Heavy With Confederate Symbols," *Atlanta Journal-Constitution*, September 4, 2015, www.ajc.com/news/state--regional-govt--politics/georgia-capitol-heavy-with-confederate-symbols/z051suE0a7bqO5cWhXlZnJ/.

15 In April 2019, Georgia Governor Brian Kemp signed SB 77, which amended existing law to expressly protect Confederate memorials (Georgia State Senate, *State Flag, Seal, and Other Symbols*). One of the bill's co-sponsors was Republican Rep. Bill Heath of the 31st district, which includes Cedartown. The Trump White

House followed suit with a June 26, 2020, Executive Order "Combatting Violence and Protecting Monuments, Memorials, and Statues" (Executive Order 13933).

16 Raphael Sassaki, "Moore on Jerusalem, Eternalism, Anarchy, and Herbie!" *Alan Moore World*, November 18, 2019; Chris Gavaler, "The Ku Klux Klan and the Birth of the Superhero," *Journal of Graphic Novels and Comics* 4, no. 2 (2012): 191–208.

17 Nguyen, "How 'Watchmen's' Misunderstanding of Vietnam Undercuts its Vision of Racism."

18 See Jonathan Gray, "*Watchmen* After the End of History."

19 *Lovecraft Country* also filmed a recreation of the Greenwood massacre in Macon, but used its Confederate monument to anchor scenes where white supremacists attack two young queer Black men; one of the show's time-traveling Black protagonists, Atticus Freeman (Jonathan Majors), comes to their defense. See Mychal Reiff-Shanks, "The Utilization of Historical Reenactment in *Lovecraft Country*," *First Thoughts on Lovecraft Country*, a special issue of *Studies in the Fantastic* 12 (Winter 2021): 55–73. The inclusion of the monument, however, dissembles its existence in the show's filming location by narratively relocating it to Oklahoma City, which doesn't include any Confederate memorials.

20 The show is based on a series of novels by Melinda Metz called *Roswell High*. The novels also inspired an earlier TV show called *Roswell* (1999–2001), which anglicized the character of Liz.

21 Sky Wu, "It Takes a Character to Make a Character—Alumni Profile, Leah Longoria '12," *The Amherst Student*, November 13, 2021, https://amherststudent.com/article/it-takes-a-character-to-make-a-character/; "Ariana Quiñonez, Writer, Mariachi, Corajuda," www.arianaquinonez.com.

22 Joe Penney, "How Medical Negligence at the US Border Killed an Immigrant Father," *The Nation*, February 25, 2020, www.thenation.com/article/world/ice-death-negligence/.

23 ACLU New Mexico, "Immigrant Rights Advocates Demand Investigation and Meeting with ICE and Elected Officials to Address Mistreatment of Gay Men and Transgender Women in Otero County Processing Center," March 22, 2019, www.aclu-nm.org/en/press-releases/immigrant-rights-advocates-demand-investigation-and-meeting-ice-and-elected-officials.

24 A prison tower can also be seen in Liz's dream sequence in S4E9.

25 Gary Farmer, "Tamara Podemski Talks with Gary Farmer about Outer Range, Reservation Dogs, Broadway, and Coroner," *Film Talk Radio*, Santa Fe, New Mexico, Augusts 23, 2021, www.youtube.com/watch?v=Lw6MH6EiujM.

26 Rachael Leishman, "Tamara Podemski Brings Change to the Western Genre as Joy in 'Outer Range,'" *The Mary Sue*, May 13, 2022, www.themarysue.com/interview-tamara-podemski-outer-range/.

27 Jeff Spry, "*Outer Range* Star Tamara Podemski Explains her Cliffhanger Ending and Season 2 Hopes," *Inverse*, May 13, 2022, www.inverse.com/entertainment/outer-range-tamara-podemski-cliffhanger-season-2.

28 Mark Blankenship, "*Outer Range*'s Tamara Podemski on Playing a Native Sheriff in a Mostly White Town," *Primetimer*, May 6, 2022, www.primetimer.com/quickhits/for-outer-ranges-tamara-podemski-its-meaningful-to-play-a-native-sheriff-in-a-mostly-white-town.

29 See Brian Davids, "'It Felt Like Old-School Hollywood': THR Presents Q&A With the Cast of 'Outer Range,'" *Hollywood Reporter*, June 16, 2022, www.hollywoodreporter.com/tv/tv-features/outer-range-cast-thr-presents-qa-1235166894/.

6. BLACK ATLANTA TV

1 Nichole Perkins, "*Greenleaf* Recap: Black and Blue," *Vulture*, June 29, 2016, www.vulture.com/2016/06/greenleaf-recap-season-1-episode-4.html.

2 Zoe Haylock, "*Being Mary Jane* Season 4, Episode 15 Recap: Feeling Hashtagged," Refinery 29, August 15, 2017, www.refinery29.com/en-us/2017/08/168147/being-mary-jane-recap-season-4-episode-15.

3 Maurice J. Hobson, *The Legend of the Black Mecca: Politics and Class in the Making of Modern Atlanta* (Chapel Hill: University of North Carolina Press, 2017), 3.

4 Ibid., 1.

5 Ibid., 3.

6 Black Atlanta TV is part of a larger group of Black shows. See Beretta E. Smith-Shomade, ed., *Watching While Black Rebooted!: The Television and Digitality of Black Audiences* (New Brunswick, NJ: Rutgers University Press, 2023); Smith-Shomade, ed., *Watching While Black: Centering the Television of Black Audiences* (New Brunswick, NJ: Rutgers University Press, 2012); and Smith-Shomade, *Pimpin' Ain't Easy: Selling Black Entertainment Television* (New York: Taylor and Francis, 2012).

7 The intergenerational HBCU address of *The Quad* was anticipated by *A Different World*. Means Coleman and Cavalcante, "Two Different Worlds," in Smith-Shomade, *Watching While Black*, 38.

8 Southern Poverty Law Center, "Flags and Other Symbols used by Far-Right Groups in Charlottesville," August 12, 2017, www.splcenter.org/hatewatch/2017/08/12/flags-and-other-symbols-used-far-right-groups-charlottesville.

9 Herman Gray, "Forward," Smith-Shomade, *Watching While Black Rebooted!*, xi. In the same volume, see Eric Pierson, "Audiences and the Televisual Slavery Narrative" (15–26); and Christine Acham, "History, Trauma, and Healing in Ava DuVernay's *13th* and *When They See Us*" (27–41).

10 Ibid., 29.

11 Dylan Rodríguez, *White Reconstruction: Domestic Warfare and the Logics of Genocide* (New York: Fordham University Press, 2020).

12 Jamel Santa Cruze Bell and Ronald L. Jackson II, eds., *Interpreting Tyler Perry: Perspectives on Race, Class, Gender, and Sexuality* (New York: Routledge, 2014); LeRhonda S. Manigault-Bryant, Tamura A. Lomax, and Carol B. Duncan, eds., *Womanist and Black Feminist Responses to Tyler Perry's Productions* (New York: Palgrave Macmillan, 2014); TreaAndrea M. Russworm, Samantha N. Sheppard,

and Karen M. Bowdre, eds., *From Madea to Media Mogul: Theorizing Tyler Perry* (Jackson: University Press of Mississippi, 2016).

13 Nikol G. Alexander-Floyd, *Re-Imagining Black Women: A Critique of Post-Feminist and Post-Racial Melodrama in Culture and Politics* (New York: New York University Press, 2021), 104.

14 Shelleen Greene, "*Tyler Perry's Too Close to Home*: Black Audiences in the Post-Network Era," in *Watching While Black Rebooted!: The Television and Digitality of Black Audiences*, ed. Beretta E. Smith-Shomade, 150–67 (New Brunswick, NJ: Rutgers University Press), 155.

15 Curtis Marez, *University Babylon*.

16 "The Most Expensive Private Jet in the World," Private-Jet.pro, https://private-jets.pro/en/guide/most-expensive-private-jet-world.html.

17 Janice D. Hamlet, *Tyler Perry: Interviews* (Jackson: University Press of Mississippi, 2019), 74.

18 Ben Brasch and J. Scott Trubey, "Tyler Perry seeks $1.8M tax break for a jet he wants to keep in Cobb," *Atlanta Journal-Constitution*, July 25, 2017, www.ajc.com/news/local/tyler-perry-seeks-tax-break-for-jet-wants-keep-cobb/iRRS3UNhsW7XuocwSAwFNK/.

19 "Flying RC Planes with Tyler Perry," www.youtube.com/watch?v=hmyi5khnzdo.

20 Pierre Bourdieu, *Distinction: A Social Critique of the Judgement of Taste* (Cambridge, MA: Harvard University Press, 1996), 218, 317.

21 Hamlet, *Tyler Perry*, 53.

22 Ibid., 8.

23 Ibid., 73–4. See also 6–7, 16.

24 Quoted by Harriette Cole, "Tyler Perry is Flying High and Enjoying the View," *AARP*, August 2, 2022, www.aarp.org/entertainment/celebrities/info-2022/tyler-perry.html.

25 Ibid., 21.

26 Ibid., 24–5.

27 "Tyler Perry Gives Exclusive Interview on *Joe Scarborough Presents*," Tyler Perry .com, May 9, 2023, https://tylerperry.com/tyler-perry-gives-exclusive-interview-on-joe-scarborough-presents/.

28 Hamlet, *Tyler Perry*, 8.

29 Nikki Finke, "Tyler Perry Fires 4 Writers for Union Activity; Atlanta Opening of Perrys New Studio Will Be Picketed; Invited Actors and Guests Being Asked Not to Attend," *Deadline*, October 2, 2008, https://deadline.com/2008/10/writers-at-tyler-perry-studio-to-take-strike-action-will-picket-grand-opening-and-ask-invited-guests-not-to-attend-7129/.

30 Aymar Jean Christian and Khadija Costly White, "One Man Hollywood: The Decline of Black Creative Production in Post Network Television," in *From Madea to Media Mogul*, 144.

31 Murali Balaji, "Tyler Perry and the Cultural Industries: New Model of Cultural Production or a Re-Version of the Old," *Interpreting Tyler Perry*, 90.

32 George Lipsitz, *How Racism Takes Place* (Philadelphia: Temple University Press, 2011), 93.
33 Henry Hollis, "Tyler Perry Donates $50K in Gift Cards to Peoplestown Community Through APD," *Atlanta Journal-Constitution*, July 17, 2020, www.ajc.com/news/tyler-perry-donates-50k-in-gift-cards-to-peoplestown-community-through-apd/Z4IV5ZOZ4FD4RKBGMU5OZ6ZC5I/.
34 "Tyler Perry's Oscar Speech Delivers a Message of Hope and a Refusal of Hate," Tyler Perry.com, https://tylerperry.com/tyler-perrys-oscar-speech-delivers-a-message-of-hope-and-a-refusal-of-hate/.
35 "Tyler Perry Studios Welcomed the Atlanta Police Department in Appreciation Luncheon," Tyler Perry.com, https://tylerperry.com/tyler-perry-studios-welcomed-the-atlanta-police-department-in-appreciation-luncheon/.
36 Catherine Hong, "Step Onto Tyler Perry's 300-Acre Production Studio," *Architectural Digest*, November 7, 2019, www.architecturaldigest.com/story/step-onto-tyler-perrys-300-acre-production-studio.
37 Kandace L. Harris, "Introduction: Being Mara," in *Representations of Black Womanhood on Television: Being Mara Brock Akil*, eds. Shauntae Brown White and Kandace L. Harris, 1–9 (Lanham, MD: Lexington Books, 2019), 1.
38 Adrienne Gaffney, "*Being Mary Jane* Creator Mara Brock Akil on Her Flawed Heroine, the Rise of Diverse TV, and Why She Hates Color-blind Casting," *Vulture*, February 3, 2015, www.vulture.com/2015/02/being-mary-jane-mara-brock-akil.html.
39 "Talks and Conversations with Mara Brock Akil," The Harry Walker Agency, www.harrywalker.com/speakers/mara-brock-akil.
40 On the show's Black feminism, see Kay Siebler, "Black Feminists in Serialized Dramas: The Gender/Sex/Race Politics of *Being Mary Jane* and *Scandal*," *Journal of Popular Film and Television* 46, no. 3 (September 10, 2019): 152–62; Kristen J. Warner, "Being Mary Jane: Cultural Specificity," in *How to Watch TV*, eds. Ethan Thompson and Jason Mittell (New York: New York University Press, 2020), 108–16; Shauntae Brown White, "'Girl, You Know I Got You': The Ideology of Sisterhood on *Being Mary Jane*," in *Representations of Black Womanhood on Television*, 64–84.
41 Natasha R. Howard, "Real, Respectable, or Both: Respectability on *Being Mary Jane* Through the Words of Mara Brock Akil," *Representations of Black Womanhood on Television*, 50.
42 Ibid., 60.
43 Quoted by Gaffney, "*Being Mary Jane* Creator Mara Brock Akil on Her Flawed Heroine, the Rise of Diverse TV, and Why She Hates Color-blind Casting."
44 Gary Klein, "Mill Valley Native Sues TV Show in Copyright Case," *Marin Independent Journal*, October 7, 2006, www.marinij.com/2006/10/07/mill-valley-native-sues-tv-show-in-copyright-case/; and Klein, "Tam High Grad, TV Networks Settle Suit," *Marin Independent Journal*, September 3, 2007, www.marinij.com/2007/09/03/tam-high-grad-tv-networks-settle-suit/.

45 "The Negro Family: The Case for National Action" (1965), informally called the Moynahan report after its author, Daniel Patrick Moynahan (Assistant Secretary of Labor under President Lyndon Johnson), aimed to convince the President that the problems faced by Black communities were not structural but caused by dysfunctional families. The report was effectively an argument against civil rights legislation since Black problems were supposedly caused by the absence of fathers and the influence of bad mothering.

46 Alexander-Floyd, *Re-Imagining Black Women,* 8–10, 12, 106, 136.

7. GENTRIFICATION TV

1 Qiana Whitted, *EC Comics: Race, Shock, and Social Protest* (New Brunswick, NJ: Rutgers University Press, 2019), 16–17.

2 Brad Curran, "Keith David Interview: *Creepshow,*" *Screen Rant,* December 9, 2021, Screen Rant, https://screenrant.com/creepshow-season-2-keith-david-interview/.

3 Research on Atlanta, for example, suggests that studios are sited in lower-priced areas where their presence increases home prices. Velma Zahirovic-Herbert and Karen M. Gibler, "The Effect of Film Production Studios on Housing Prices in Atlanta, the Hollywood of the South," *Urban Studies* 59, no. 4 (July 29, 2021), https://journals.sagepub.com/doi/abs/10.1177/00420980211024156.

4 Mayer, *Almost Hollywood, Nearly New Orleans,* 44. Helen Morgan Parmett similarly argues that the show *Portlandia* helped drive gentrification in Portland's historic Black neighborhoods. See her essay, "Site-Specific Television as Urban Renewal: Or, How Portland became *Portlandia,*" *International Journal of Cultural Studies* 21, no. 1 (2018): 42–56.

5 Mayer, *Almost Hollywood, Nearly New Orleans,* 55–59.

6 Ian R. Cook and Ishan Ashutosh, "Television Drama and the Urban Diegesis: Portraying Albuquerque in *Breaking Bad,*" *Urban Geography* 39, no. 5 (2018): 746–62; Alexander Gutzmer, *TV Shows and Nonplace: Why* The Sopranos, Breaking Bad *and Co. Love the Periphery* (London: Taylor & Francis, 2023), 69–71.

7 Trilith Studios website, www.trilithstudios.com.

8 Pat Saperstein, "Is Trilith, a 'Company Town' for Marvel's Georgia Production Workers, the Template of the Future?," *Variety,* March 10, 2022, https://variety.com/2022/film/news/marvel-trilith-town-georgia-1235194027/.

9 Briana Sacks, "Trilith Promised a New Model for Hollywood in the Atlanta Suburbs, But Black Residents and Former Employees Say Racism Lies Just Below the Surface," *BuzzFeed News,* July 18, 2022, www.buzzfeednews.com/article/briannasacks/trilith-georgia-studios-town-race.

10 Rodney Ho, "Black Residents of Trilith Sue Studio, Development for Discrimination," *Atlanta Journal-Constitution,* November 17, 2022, www.ajc.com/life/radiotvtalk-blog/black-residents-of-trilith-sue-studio-development-for-discrimination/4ALI4D33UFFQFLUXQTHN5M3T64/.

11 "Site History," promotional video, Assembly Atlanta website, https://assemblyatlanta.com.

12 Angelina Velasquez, "Tyler Perry Buys More Land to Expand His Atlanta Studio and Develop an Entertainment Complex That Boasts Theaters and Restaurants," *Atlanta Black Star*, January 9, 2023, https://atlantablackstar.com/2023/01/09/tyler-perry-completes-8m-purchase-of-additional-acres-to-expand-his-atlanta-studio/.
13 "For Sale: Homes From 'The Walking Dead' Set," *Fox 5 Atlanta*, May 20, 2022, www.fox5atlanta.com/news/for-sale-homes-from-the-walking-dead-set.
14 Scott Tigchelaar, "Georgia Department of Economic Development," www.georgia.org/node/45861; Bouckaert Farm website, "Productions at Bouckaert Farm," https://bouckaertfarm.com/productions/.
15 Velma Zahirovic-Herbert and Karen M. Gibler, "The Effect of Film Production Studios on Housing Prices in Atlanta," *Urban Studies* 59, no. 4 (2021): 771–88.
16 See Immergluck, *Red Hot City*.
17 Jenny Fuster, "Atlanta Studio Sparks Protests for Plan to Clearcut 200-Acre Forest for More Soundstages," *TheWrap*, July 2021, https://www.thewrap.com/atlanta-blackhall-studio-protests-200-acre-forest-soundstages/
18 Mesa del Sol website, www.mesadelsolnm.com.
19 "Trilith Named Mixed-Use Community of the Year," *The Atlanta Real Estate Forum*, November 9, 2021, www.atlantarealestateforum.com/trilith-development-mixed-use-coy/.
20 "About Senoia," Gin Property Group website, https://ginproperty.com/about-senoia/.
21 George J. Sánchez, "Why Are Multiracial Communities So Dangerous? A Comparative Look at Hawai'i: Cape Town, South Africa; and Boyle Heights, California," *Pacific Historical Review* 86, no. 1 (2007): 166.
22 Monarch Studios Los Angeles website, https://monarchstudiosla.com.
23 Pollution Studios website, https://pollutionstudios.com.
24 Kean O'Brian, Leonardo Vilchis, and Croina Maritescu, "Boyle Heights and the Fight against Gentrification as State Violence," *American Quarterly* 71, no. 2 (June 2019), 389.
25 Anastasia Baginski and Chris Malcolm, "Gentrification and the Aesthetics of Displacement," *Field: A Journal of Socially Engaged Art Criticism* 14 (Fall 2019), http://field-journal.com/issue-14/gentrification-and-the-aesthetics-of-displacement.
26 Ultra-red, "*Desarmando Desarrollismo*: Listening to Anti-Gentrification in Boyle Heights," *Field: A Journal of Socially Engaged Art Criticism* 14 (Fall 2019), http://field-journal.com/issue-14/desarmando-desarrollismo.
27 O'Brian, Vilchis, and Maritescu, "Boyle Heights and the Fight against Gentrification as State Violence," 393.
28 Pulido, Barraclough, and Cheng, *A People's Guide to Los Angeles*, 97.
29 "As Rents Soar in L.A, Even Boyle Heights' Mariachis Sing the Blues," *The Los Angeles Times*, September 9, 2017, www.latimes.com/local/lanow/la-me-boyle-heights-musicians-gentrification-20170909-htmlstory.html.
30 In their content analysis of seven seasons of the show, Jacob S. Turner and Lisa G. Perk found that humans were responsible for 80 (65.6%) of the killings compared to only 20 (16.4%) for zombies/walkers and 22 (18.0%) resulting from

other causes. See "White Men Holding on for Dear Life and Taking It: A Content Analysis of the Gender and Race of the Victims and Killers in *The Walking Dead*," *Sex Roles* 81 (2019): 660.

31 Lauren O'Mahony, Melissa Merchant, and Simon Order, "NecroPolitics in a Post-Apocalyptic Zombie Diaspora: The Case of AMC's *The Walking Dead*," *Journal of Postcolonial Writing* 57, no. 1 (2021): 89–103.

32 On the show's idealization of patriarchy, see John Green and Michaela D. E. Meyer, "The Walking (Gendered) Dead: A Feminist Rhetorical Critique of Zombie Apocalypse Television Narrative," *Ohio Communication Journal* 52 (October 2014): 64–74; Katherine Sugg, "*The Walking Dead*: Late Liberalism and Masculine Subjection in Apocalypse Fictions," *Journal of American Studies* 49, no. 4 (2015): 793–811; and Zach Finch, "The Walking Dead and Gendering Austerity," in *Gender and Austerity in Popular Culture: Femininity, Masculinity and Recession in Film and Television*, eds. Helen Davies and Claire O'Callaghan (New York: I. B. Tauris, 2017), 135–52.

33 Quoted by Nick Schager, "The Gloriously Bizarre Genius of Benny Safdie," *The Daily Beast*, December 4, 2023, www.thedailybeast.com/obsessed/benny-safdie-interview-on-the-bizarre-genius-of-the-curse.

34 Quoted by Ben Travers, "Nathan Fielder and Benny Safdie Nearly Talked Themselves Out of Making 'The Curse,'" *Indiewire*, December 4, 2023, www.indiewire.com/features/interviews/nathan-fielder-benny-safdie-the-curse-interview-1234931493/.

35 I use "Chicana" here because the character expressly identifies as a "girl" in season 3, episode 5, although the actor who plays the character, Ser Anzoategui, identifies as non-binary.

36 Jeff Goldsmith, "*Vida* Q & A—Tanya Saracho," *The Q&A with Jeff Goldsmith*, 2019, www.mixcloud.com/theqawithjeffgoldsmith/vida-qa-tanya-saracho/.

37 During the filming of season 1, BHAAD prevented *Vida* from shooting in Mariachi Plaza but the show filmed there for its third-season finale. Alejandra Reyes-Velarde, "Drama on Set and Off; 'Vida,' a TV Show About Gentrification, Faces Protests by Some Who Say it is Doing Just That," *Los Angeles Times*, September 1, 2018, A.1.

38 See Carlos Jimenez and Alfredo Huante, "Home in *Vida* and *Gentefied*: The Politics of Representation in Gente-fication Narratives," *Aztlan: A Journal of Chicano Studies* 48, no. 1 (Spring 2023): 38.

39 On gentrification and the police in Boyle Heights, see O'Brian, Vilchis, and Maritescu, 392.

40 Yvonne Villarreal, "Netflix Tackles Gentrification; 'Gentefied' Debuts Amid Concern It May Contribute to Problem in Boyle Heights," *Los Angeles Times*, February 2020, E1.

41 Anna Marta Marini, "*Gentefied* and the Representation of Gentrification Related Latinx Conflicts," *JAm It!* 4 (May 2021): 44–5.

42 Maria Elana Fernandez, "They Felt Like Hollywood Outsiders. Then They Got a Netflix Show," *Vulture*, February 21, 2020, www.vulture.com/2020/02/gentefied-netflix-creators.html.

43 Sugg, *"The Walking Dead."*
44 Jana Schmieding (@janaunplgd), Instagram post, www.instagram.com/p/CNv_cCCLvLK/; https://www.instagram.com/p/CCtflfzloNy/.
45 Weshoyot Alvitre (@weshoyot), X (formerly Twitter), February 9, 2020, https://twitter.com/weshoyot/status/1226577767854157824.
46 Quoted by Sandra Hale Schulman, "Art Installations Remind LA Residents They are on 'TONGVALAND,'" *Indian Country Today*, August 30, 2021, https://ictnews.org/news/art-installations-remind-la-residents-theyre-on-tongvaland.
47 Weshoyot Alvitre (@weshoyot), Instagram post, October 10, 2022, www.instagram.com/p/Cjid-NhJnmT/.

AFTERWORD

1 Sarah Arnold, "Netflix and the Myth of Choice/Participation/Autonomy," in *The Netflix Effect: Technology and Entertainment in the 21st Century*, ed. Kevin McDonald and Daniel Smith-Rowsey, 49–62 (New York: Bloomsbury Publishing, 2018), 49–51. See also Mareike Jenner, *Netflix and the Reinvention of Television* (New York: Palgrave Macmillan, 2018), 162.
2 Lagerwey and Nygaard first developed their claims over the course of three essays in *Flow: A Critical Forum on Media and Culture*: "Liberal Women, Mental Illness, and Precarious Whiteness in Trump's America" (November 27, 2017, www.flowjournal.org/2017/11/whiteness-in-trumps-america/); "Broad City's Affable Critique and the Racial Discourses of Girlfriendship" (March 26, 2018, www.flowjournal.org/2018/03/broad-city-critique-girlfriendship/); and "Sitcom in the Trump Era: Racially Diverse Utopias and White Dystopias" (May 29, 2018, www.flowjournal.org/2018/05/sitcom-in-the-trump-era/).
3 Lagerwey and Nygaard, *Horrible White People: Gender, Genre, and Television's Precarious Whiteness* (New York: New York University Press, 2020), 9–10.
4 Ruha Benjamin, *Race After Technology: Abolitionist Tools for the New Jim Code* (New York: Polity, 2019), 18. See also Safiya Umoja Noble, *Algorithms of Oppression: How Search Engines Reinforce Racism* (New York: New York University Press, 2018).
5 Benjamin, *Race After Technology*, 5–6.
6 Zachary Snider, "The Cognitive Psychological Effects of Binge-Watching," *The Netflix Effect*, 117–19.
7 Jenner, *Netflix and the Reinvention of Television*, 119–34, 161–62.
8 See Miriam Hansen, *Babel and Babylon: Spectatorship in American Silent Film* (Cambridge, MA: Harvard University Press, 1994); and Marez, *Drug Wars*.
9 On representations of AI in film and television, see Isabella Herman, "Artificial Intelligence in Fiction: Between Narratives and Metaphors," *AI and Society* 38 (2023): 319–29. See also the cluster of essays on AI in *The Palgrave Handbook of Posthumanism in Film and Television*, eds. Michael Hauskeller, Thomas D. Philbeck, and Curtis D. Carbonell (New York: Palgrave Macmillan, 2015): Kevin LaGrandeur, "Androids and the Posthuman in Television and Film," 111–19; Sherryl

Vint, "'Change for the Machines'? Posthumanism as Digital Sentience," 120–29; Jeff Menne and Jay Clayton, "Alive in the Net," 130–40; Dónal P. O'Mathúna, "Autonomous Fighting Machines: Narratives and Ethics," 141–52.

10 Quoted in Alan Sepinwall, "'There's Power in Laughing at the Pain': 'Brockmire' Creator on Series' American Nightmare," *Rolling Stone*, May 5, 2020, www.rollingstone.com/tv-movies/tv-movie-features/brockmire-finale-creator-church-cooper-interview-990616/.

11 See Benjamin, *Race After Technology*; Herman, "Artificial Intelligence in Fiction," and Noble, *Algorithms of Oppression*.

12 Nantheera Annatrasirchai and David Bull, "Artificial Intelligence in the Creative Industries: A Review," *Artificial Intelligence Review* 55 (2022): 589–656. One possible exception would be reality TV shows, where AI would replace reality TV extras and writers, among the cheapest labor pools in the industry.

13 Katie Kilkenny, "Tyler Perry Puts $800M Studio Expansion on Hold After Seeing OpenAI's Sora: 'Jobs Are Going to Be Lost,'" *The Hollywood Reporter*, February 22, 2024, www.hollywoodreporter.com/business/business-news/tyler-perry-ai-alarm-1235833276/.

14 "The Ultimate Platform for OpenAI Sora Generated Videos and Prompts," SoraHub website, https://sorahub.video.

15 Elijah Clark, "Tyler Perry Warns of AI Threat After Sora Debut Halts an $800 Million Studio Expansion," *Forbes*, February 23, 2024, www.forbes.com/sites/elijahclark/2024/02/23/tyler-perry-warns-of-ai-threat-to-jobs-after-viewing-openai-sora/?sh=10e7588d7071.

16 Ann-Laure Ligozat, Julien Lefèvre, Aurélie Bugeau, and Jacques Combaz, "Unraveling the Hidden Environmental Impacts of AI Solutions for Environment Life Cycle Assessment of AI Solutions," *arXiv*, April 12, 2022, https://arxiv.org/abs/2110.11822v2; Scott Robbins and Aimee van Wynsberghe, "Our New Artificial Intelligence Infrastructure: Becoming Locked into an Unsustainable Future," *Sustainability* 14 (2024): 1–11, www.mdpi.com/2071-1050/14/8/4829.

BIBLIOGRAPHY

Abid-Kons, Sukainah. "Fight Against 'Cop City' Continues at the 'Community in Weelaunee' Summit." *Georgia Voice*, May 5, 2022. https://thegavoice.com/community/fight-against-cop-city-continues-at-the-community-in-weelaunee-summit/.

ACLU New Mexico. "Immigrant Rights Advocates Demand Investigation and Meeting with ICE and Elected Officials to Address Mistreatment of Gay Men and Transgender Women in Otero County Processing Center." March 22, 2019. www.aclu-nm.org/en/press-releases/immigrant-rights-advocates-demand-investigation-and-meeting-ice-and-elected-officials.

Albuquerque Development Commission. "LEDA-20-7: Netflix Studio Project" (report). November 23, 2020. www.cabq.gov/mra/documents/case-2020-10-leda-20-7-staff-analysis-netflix-studios.pdf.

Albuquerque Police Department. Facebook. December 13, 2016. www.facebook.com/abqpolice/photos/a.482215295149670/1216874778350381/?type=3.

Alexander-Floyd, Nikol G. *Re-Imagining Black Women: A Critique of Post-Feminist and Post-Racial Melodrama in Culture and Politics*. New York: New York University Press, 2021.

All Monuments Must Fall: A Collaboratively Produced Syllabus. July 2020. https://allmonumentsmustfall.com.

Alvitre, Weshoyot (@wweshoyot). Instagram, October 10, 2022. www.instagram.com/p/Cjid-NhJnmT/.

Amiker, Francesca. "A Look at the Studio and #1 Job Behind Jumanji, Venom, Godzilla and More." *11 Alive News*, Atlanta, April 27, 2018. www.11alive.com/article/news/a-look-at-the-studio-and-1-job-behind-jumanji-venom-godzilla-and-more/85-546478414.

Annatrasirchai, Nantheera, and David Bull. "Artificial Intelligence in the Creative Industries: A Review." *Artificial Intelligence Review* 55 (2022): 589–656.

Arnold, Sarah. "Netflix and the Myth of Choice/Participation/Autonomy." In *The Netflix Effect: Technology and Entertainment in the 21st Century*, ed. Kevin McDonald and Daniel Smith-Rowsey, 49–62. New York: Bloomsbury Publishing, 2018.

Atkin, Hillary. "Swarm Co-Creator Janine Nabers on Working With Donald Glover, Billie Eilish, and Fellow Emmy Nominee Dominique Fishback." *Above the Line*, August 14, 2022. https://abovetheline.com/2023/08/14/swarm-janine-nabers-interview-donald-glover-billie-eilish-dominique-fishback/.

Aurthur, Kate. "From Daniel's Death to Native Humor, 'Reservation Dogs' Showrunner Sterlin Harjo Answers Our Burning Questions." *Variety*, June 15, 2022. https:

//variety.com/2022/tv/news/reservation-dogs-sterlin-harjo-daniel-death-native-humor-1235294333/.

Auslander, Mark, and Avis E. Williams. "What these Trees have Seen: Slavery, Post-Slavery, and Anti-Blackness in the South River (Weelaunee) Forest Zone." Mark Auslander: New Visons on Museums, Community Engagement, Art, and Science in the Public Interest (website). April 23, 2022. https://markauslander.com/2022/08/04/what-these-trees-have-seen-slavery-post-slavery-and-anti-blackness-in-the-south-river-welaunee-forest-zone/.

Báez, Jillian. "Television for All Women? Watching Lifetime's *Devious Maids*." In *Cupcakes, Pinterest, and Lady Porn: Feminized Popular Culture in the Early Twenty-First Century*, ed. Elana Levine, 51–70. Chicago: University of Illinois Press, 2015.

Baginski, Anastasia, and Chris Malcolm. "Gentrification and the Aesthetics of Displacement." *Field: A Journal of Socially Engaged Art Criticism* (Fall, 2019). http://field-journal.com/issue-14/gentrification-and-the-aesthetics-of-displacement.

Banks, Jonathan, Bryan Cranston, Vince Gilligan, Gennifer Hutchinson, and Michael Slovis. "DVD Commentary on 'No Mas.'" *Breaking Bad*, S3E1. AMC, 2010.

Banks, Miranda, Bridget Conor, and Vicki Mayer, eds. *Production Studies, The Sequel!: Cultural Studies of Global Media Industries*. New York: Routledge, 2015.

Barton, Kelsie. "Here's an Exclusive Peek at Future Plans for ATL's Bouckaert Farm." *The Atlantan*, March 29, 2021. https://atlantanmagazine.com/exclusive-peek-at-future-plans-atl-bouckaert-farm.

Bell, Jamel Santa Cruze and Ronald L. Jackson II, eds. *Interpreting Tyler Perry: Perspectives on Race, Class, Gender, and Sexuality*. New York: Routledge, 2014.

Benjamin, Ruha. *Race After Technology: Abolitionist Tools for the New Jim Code*. New York: Polity, 2019.

Bennett, James T. *They Play, You Pay: Why Taxpayers Build Ballparks, Stadiums, and Arenas for Billionaire Owners and Millionaire Players*. New York: Springer, 2012.

Blankenship, Mark. "*Outer Range*'s Tamara Podemski on Playing a Native Sheriff in a Mostly White Town." *Primetimer*, May 6, 2022. www.primetimer.com/quickhits/for-outer-ranges-tamara-podemski-its-meaningful-to-play-a-native-sheriff-in-a-mostly-white-town.

Bodroghkozy, Aniko. "From Civil Rights to Unite the Right: What the Photographs Say." *Boston Globe*, August 10, 2022. www.bostonglobe.com/2022/08/10/opinion/civil-rights-unite-right-what-photographs-say/.

Bourdieu, Pierre. *Distinction: A Social Critique of the Judgement of Taste*. Cambridge, MA: Harvard University Press, 1996.

Bradbury, John Charles. "Do Movie Production Incentives Generate Economic Development?" *Contemporary Economic Policy* 38, no. 2 (August 2020): 327–342.

———. "Film Tax Credits and the Economic Impact of the Film Industry on Georgia's Economy." *Bagwell Center Policy Brief*. Kennesaw State University, July 2019. https://papers.ssrn.com/sol3/papers.cfm?abstract_id=3407921.

Bradley, Bill. "What Happened to Skinny Pete In 'El Camino'? This Real-Life Police Officer Knows." *Huffington Post*, October 17, 2019. www.huffpost.com/entry/what-happened-skinny-pete-el-camino-police_n_5da109f4e4b06ddfc519f0e1.

Brasch, Ben, and J. Scott Trubey. "Tyler Perry seeks $1.8M tax break for a jet he wants to keep in Cobb." *Atlanta Journal-Constitution*, July 25, 2017. www.ajc.com/news/local/tyler-perry-seeks-tax-break-for-jet-wants-keep-cobb/iRRS3UNhsW7Xu0cwSAwFNK/.

Bret, Jennifer. "Old Atlanta Prison Farm a Favorite with Filmmakers." *Atlanta Journal-Constitution*, October 31, 2014. www.ajc.com/blog/buzz/old-atlanta-prison-farm-favorite-with-filmmakers/yHzdSgh4ywGlzFaxjokqdL/.

Brown, Shauntae White, and Kandace L. Harris, eds. *Representations of Black Womanhood on Television: Being Mara Brock Akil*. Lanham, MD: Lexington Books, 2019.

Bruns, Adam. "Electric Owl Studios Comes In For a Landing." *Site Selection Magazine*, March 2022. https://siteselection.com/investor-watch/electric-owl-studios-comes-in-for-a-landing.cfm.

Burroughs, Benjamin, Benjamin J. Morse, Travis Snow, and Michael Carmona. "The Masks We Wear: *Watchmen*, Infrastructural Racism, and Anonymity." *Television and New Media* 23, no. 3 (2023): 247–263.

Button, Patrick. "Can Tax Incentives Create a Local Film Industry? Evidence from Louisiana and New Mexico." *Journal of Urban Affairs* 43, no. 5 (2021): 658–684.

Caldwell, John Thornton. *Production Culture: Industrial Reflexivity and Critical Practice in Film and Television*. Durham, NC: Duke University Press, 2008.

Camp, Jordan T. and Christina Heatherton. "Riots in the Master's Hall: Racism, Nationalism, and the Crisis of US Hegemony." In *Racism, Violence and Harm: Ideology, Media, and Resistance*, ed. Monish Bhatia, Scott Poynting, and Waqas Tufail, 225–242. New York: Palgrave Macmillan, 2023.

Campbell, Faye. "Multifaceted Meanings of Monuments Continue to Evolve." *Newnan Times-Herald*, June 20, 2020. www.times-herald.com/news/local/multi-faceted-meanings-of-monuments-continue-to-evolve/article_da29e9eb-ae9e-51d8-8f1d-48b0ae6f33f0.html.

Castañeda, Mari Paredes. "Television Set Production at the US–Mexico Border: Trade Policy and Advanced Electronics for the Global Market." In *Critical Cultural Policy: A Reader*, ed. Justin Lewis and Toby Miller, 272–281. Malden, MA: Blackwell, 2002.

Chait, Jonathan. "American Death Cult: Why has the Republican Response to the Pandemic Been so Mind-Bogglingly Disastrous?" *New York Magazine*, July 20, 2020. https://nymag.com/intelligencer/2020/07/republican-response-coronavirus.html.

Chin, Derek. "Profile of New Mexico Prison Population." New Mexico Sentencing Commission, December 2021. https://nmsc.unm.edu/reports/2021/profile-of-new-mexico-prison-population.pdf.

Chin, Helen, and Onyx Jones. "Reparations: A Journey Toward Repair." City of Culver City, December 14, 2022. https://oag.ca.gov/system/files/media/task-force-city-culver-city-presentation-12142022-12152022.pdf.

City of Albuquerque. "City Making Major Upgrades, Improving Access to Mesa Del Sol." 2020. www.cabq.gov/municipaldevelopment/news/city-making-major-upgrades-improving-access-to-mesa-del-sol.

Clark, Elijah. "Tyler Perry Warns of AI Threat After Sora Debut Halts an $800 Million Studio Expansion." *Forbes*, February 23, 2024. www.forbes.com/sites/elijahclark/2024/02/23/tyler-perry-warns-of-ai-threat-to-jobs-after-viewing-openai-sora/?sh=10e7588d7071.

Classen, Steven D. *Watching Jim Crow: The Struggles over Mississippi TV, 1955–1969*. Durham, NC: Duke University Press, 2004.

Cole, Harriette. "Tyler Perry is Flying High and Enjoying the View." *AARP*, August 2, 2022, www.aarp.org/entertainment/celebrities/info-2022/tyler-perry.html.

Coleman, Madeline Leung. "Donald Glover Hoped *Swarm* Would Make You Uncomfortable." *Vulture*, March 28, 2023. www.vulture.com/article/donald-glover-swarm-interview-dre-dominique-fishback.html.

Coleman, Robin R. Means, and Andre M. Cavalcante. "Two Different Worlds: Television as a Producer's Medium." In *Watching While Black: Centering the Television of Black Audiences*, ed. Beretta E. Smith-Shomade, 33–48. New Brunswick, NJ: Rutgers University Press, 2012.

Color of Change and the USC Annenberg Norman Lear Center. *Normalizing Injustice: The Dangerous Misrepresentations that Define Television's Scripted Crime Genre*. Study. January 2020. https://colorofchange.org/press_release/normalizing-injustice-new-landmark-study-by-color-of-change-reveals-how-crime-tv-shows-distort-understanding-of-race-and-the-criminal-justice-system/.

Cook, Ian R., and Ishan Ashutosh. "Television Drama and the Urban Diegesis: Portraying Albuquerque in *Breaking Bad*." *Urban Geography* 39, no. 5 (2018): 746–62.

Cooper, Bertrand. "Who Actually Gets to Create Black Pop Culture?" *Current Affairs*, July 25, 2021. www.currentaffairs.org/2021/07/who-actually-gets-to-create-black-pop-culture.

Couch, Aaron. "New Mexico Governor Signs 'Breaking Bad' TV, Film Subsidy Bill into Law." *Hollywood Reporter*, April 4, 2013. www.hollywoodreporter.com/news/politics-news/new-mexico-governor-signs-breaking-433168/.

Covington, Floyd C. "The Negro Invades Hollywood." *Opportunity* 7, no. 4 (April 1929): 111–113.

Cowie, Jefferson. *Capital Moves: RCA's Seventy-Year Quest for Cheap Labor*. Ithaca, NY: Cornell University Press, 2001.

Curran, Brad. "Keith David Interview: *Creepshow*." *Screen Rant*, December 9, 2021. https://screenrant.com/creepshow-season-2-keith-david-interview/.

Curtin, Michael, and Kevin Sanson, eds. *Precarious Creativity: Global Media, Local Labor*. Oakland: University of California Press, 2016.

D'Alessandro, Anthony. "Emmys: 'Devious Maids' Second Life After its Death at ABC; Creator Marc Cherry Q&A." *Deadline*, June 7, 2014. https://deadline.com/2014/06/emmys-devious-maids-second-life-after-its-death-at-abc-creator-marc-cherry-qa-741969/.

Daniel, Victor. "Hart Family Parents Killed 6 Children in Murder-Suicide, Jury Decides." *New York Times*, April 5, 2019. www.nytimes.com/2019/04/05/us/hart-family-murder-suicide.html.

Danielson, Michael N. *Home Team: Professional Sports and the American Metropolis*. Princeton, NJ: Princeton University Press, 1997.

Davids, Brian. "'It Felt Like Old-School Hollywood': THR Presents Q&A With the Cast of 'Outer Range.'" *Hollywood Reporter*, June 16, 2022. www.hollywoodreporter.com/tv/tv-features/outer-range-cast-thr-presents-qa-1235166894/.

Delaney, Kevin J., and Rick Eckstein. *Public Dollars, Private Stadiums: The Battle over Building Sports Stadiums*. New Brunswick, NJ: Rutgers University Press, 2003.

DiRienzo, Rob. "Anonymous Activist Claims Responsibility for Blaze at DeKalb Film Studio: 'May This be a Warning to Them.'" *Fox 5 Atlanta*, October 21, 2022. www.fox5atlanta.com/news/anonymous-activist-claims-responsibility-for-blaze-at-dekalb-film-studio-may-this-be-a-warning-to-them.

Donovan, Thomas, J. "Families of Three Inmates Who Died at Gwinnett Jail Push for Changes." *Atlanta Journal-Constitution*, October 15, 2022. www.ajc.com/news/families-of-3-inmates-who-died-at-gwinnett-jail-push-for-changes/MNHK6UJXD5BOZG4H5PGH65NYZQ/.

Dorr, Kirstie. *On Site, In Sound: Performance Geographies in Latina/o America*. Durham, NC: Duke University Press, 2018.

Drury, Sharareh. "'Walking Dead' Showrunner on How the Pandemic Impacts its Final Season." *Hollywood Reporter*, August 13, 2021. www.hollywoodreporter.com/tv/tv-news/walking-dead-final-season-preview-angela-kang-1234995522/.

Eady, Alyse. "Georgia's Tire Problem." *Good Day Atlanta*, Fox 5, June 29, 2018. www.youtube.com/watch?v=65krwSKXdRE.

Echols, Jacqueline. "Environmental Justice in the South River Watershed." Interview by Sorrel Inman. *Mergoat Magazine*, Spring 2023, 54–55. https://issuu.com/mergoat/docs/atl_section_-_blue_hollers_-_digital/48.

Edens, Chief J. E. "Senoia Police Department Memorandum," July 8, 2019. www.senoia.com/sites/default/files/fileattachments/city_council/meeting/4411/vehicle_purchase_request.pdf.

Estep, Tyler. "Officials: 'Historic' Film Studio Expansion to Bring 2,400 Jobs to South DeKalb, Development Authority Grants $68 Million Incentive Package." *Atlanta Journal-Constitution*, April 14, 2022. www.ajc.com/neighborhoods/dekalb/officials-historic-film-studio-expansion-to-bring-2400-jobs-to-south-dekalb/W6BZIEI2SFEC5IIGMRSN45KKJY/.

Evans, Beau. "Kemp Signs 2022 Georgia Budget, Adds Back Most Schol Funds Cut in Pandemic." *GPB News*, May 11, 2021. www.gpb.org/news/2021/05/11/kemp-signs-2022-georgia-budget-adds-back-most-school-funds-cut-in-pandemic.

Farmer, Gary. "Tamara Podemski Talks with Gary Farmer about Outer Range, Reservation Dogs, Broadway, and Coroner." *Film Talk Radio*, Santa Fe, New Mexico, August 23, 2021. www.youtube.com/watch?v=Lw6MH6EiujM.

Fassler, Ella. "Activists Occupying The Woods to Block 'Cop City' Face Terrorism Charges." *Vice*, December 21, 2022. www.vice.com/amp/en/article/xgy9yk/activists-occupying-the-woods-to-block-cop-city-face-terrorism-charges.
Faye, Jennifer. *Inhospitable World: Cinema in the Time of the Anthropocene*. New York: Oxford University Press, 2018.
Fernandez, Maria Elena. "They Felt Like Outsiders. Then They Got a Netflix Show. How Two First-Generation Artists Made their Bilingual Love Letter, *Gentefied*." *Vulture*, February, 21, 2020. www.vulture.com/2020/02/gentefied-netflix-creators.html.
Film New Mexico. "Incentives." https://nmfilm.com/why-new-mexico/incentives-2.
———. "New Mexico Tax and Revenue Payout." https://nmfilm.com/why-new-mexico/incentives-2.
———. "Production Resources: State Owned Buildings." https://nmfilm.com/filmmaker-resources/permits-procedures/state-owned-buildings.
FilmLA. *2019 Television Report*. Hollywood: FilmL.A., Inc., 2019.
Finch, Zach. "The Walking Dead and Gendering Austerity." In *Gender and Austerity in Popular Culture: Femininity, Masculinity and Recession in Film and Television*, ed. Helen Davies and Claire O'Callaghan, 135–152. New York: I. B. Tauris, 2017.
Finke, Nikki. "Tyler Perry Fires 4 Writers for Union Activity; Atlanta Opening of Perry's New Studio Will Be Picketed; Invited Actors and Guests Being Asked Not to Attend." *Deadline*, October 2, 2008. https://deadline.com/2008/10/writers-at-tyler-perry-studio-to-take-strike-action-will-picket-grand-opening-and-ask-invited-guests-not-to-attend-7129/.
Fleming, Mike, Jr. "Netflix Commits $1 Billion More In New Mexico Production Funding As It Expands ABQ Studios; 'Stranger Things' Joins List Of Albuquerque-Set Shows." *Deadline*, November 23, 2020. https://deadline.com/2020/11/netflix-billion-dollar-production-commitment-new-mexico-abq-studios-stranger-things-1234620435/.
Flook, Ray. "Watchmen: Damon Lindelof Knows What His Episode 10 Title Would Be." Bleeding Cool. June 5, 2020, https://bleedingcool.com/tv/watchmen-damon-lindelof-know-what-his-episode-10-title-would-be/.
Fortmueller, Kate. *Hollywood Shutdown: Production, Distribution, and Exhibition in the Time of COVID*. Austin: University of Texas Press, 2021.
Friends of Decatur Cemetery. "Lives That Made Our City: Decatur Cemetery Walking Tour." Decatur, GA: n.d. www.decaturga.com/media/21356.
Fuster, Jenny. "Atlanta Studio Sparks Protests for Plan to Clearcut 200-Acre Forest for More Soundstages." *TheWrap*, July 2021. www.thewrap.com/atlanta-blackhall-studio-protests-200-acre-forest-soundstages/.
Gaffney, Adrienne. "*Being Mary Jane* Creator Mara Brock Akil on Her Flawed Heroine, the Rise of Diverse TV, and Why She Hates Color-blind Casting." *Vulture*, February 3, 2015. www.vulture.com/2015/02/being-mary-jane-mara-brock-akil.html.
Gao, Max. "A TV Drama in ICE Custody." *Los Angeles Times*, February 8, 2022, E1, E6.
Gates, Raquel J. *Double Negative: The Black Image and Popular Culture*. Durham, NC: Duke University Press, 2018.

Gavaler, Chris. "The Ku Klux Klan and the Birth of the Superhero." *Journal of Graphic Novels and Comics* 4, no. 2 (2012): 191–208.
Gillespie, Michael Boyce. "Thinking about *Watchmen*: with Jonathan W. Gray, Rebecca A. Wanzo, and Kristen J. Warner." *Film Quarterly* 73, no. 4 (2002): 50–60.
Gilmore, Ruth Wilson. *Golden Gulag: Prisons, Surplus, Crisis, and Opposition to Globalization*. Berkeley: University of California Press, 2007.
Ginsburg, Faye, Brian Larkin, and Lilia Abu-Lughod, eds. *Media Worlds: Anthropology on New Terrain*. Berkeley: University of California, 2002.
Gleich, Joshua, and Lawrence Webb, eds. *Hollywood on Location: An Industry History*. New Brunswick, NJ: Rutgers University Press, 2019.
Golden, Jeremy A., Karen K. Wong, Christine M. Szablewski, et al. "Characteristics and Clinical Outcomes of Adult Patients with Covid-19—Georgia March 2020." *Morbidity and Mortality Weekly Report*, April 29, 2020. www.cdc.gov/mmwr/volumes/69/wr/mm6918e1.htm.
Goldsmith, Jeff. "*Vida* Q & A—Tanya Saracho." *The Q&A with Jeff Goldsmith*, 2019. www.mixcloud.com/theqawithjeffgoldsmith/vida-qa-tanya-saracho/.
Gravel, Lucia. "Forest Defenders Continue Fight Against 'Cop City.'" *The Southerner*, Midtown High School, Atlanta, October 19, 2022. https://thesoutherneronline.com/90048/news/forest-defenders-continue-fight-against-cop-city/#modal-photo.
Gray, Jonathan W. "*Watchmen* After the End of History." *ASAP Journal*, February 3, 2020. https://asapjournal.com/feature/watchmen-after-the-end-of-history-race-redemption-and-the-end-of-the-world-jonathan-w-gray/.
Green, John, and Michaela D. E. Meyer. "The Walking (Gendered) Dead: A Feminist Rhetorical Critique of Zombie Apocalypse Television Narrative." *Ohio Communication Journal* 52 (October 2014): 64–74.
Gregg, John. "Gary Johnson says tax breaks made New Mexico the 'Second Hollywood.'" *Politifact*. July 15, 2011. www.politifact.com/factchecks/2011/jul/15/gary-johnson/gary-johnson-says-tax-breaks-made-new-mexico-secon/.
Gray, Herman. *Cultural Moves: African Americans and the Politics of Representation*. Berkeley: University of California Press, 2005.
———. *Watching Race: Television and the Struggle for Blackness*. Minneapolis: University of Minnesota Press, 2004.
Gutzmer, Alexander. *TV Shows and Nonplace: Why the Sopranos, Breaking Bad and Co. Love the Periphery*. London: Taylor & Francis, 2023.
Gwinnett Film. "Location Search Results: Prisons/Jails," December 8, 2024. https://gwinnett.locationshub.com/search_results.aspx?search=&search_mode=and&ctry=US&state=GA&city=&bor=&cat=989&subcat=989&style=&lname=&lid=&int=&sort=date&view=16.
Haggins, Bambi. *Laughing Mad: The Black Comic Persona in Post-Soul America*. New Brunswick, NJ: Rutgers University Press, 2007.
Hamlet, Janice D. *Tyler Perry: Interviews*. Jackson: University Press of Mississippi, 2019.
Hansen, Miriam. *Babel and Babylon: Spectatorship in American Silent Film*. Cambridge, MA: Harvard University Press, 1994.

Harper, Jordan, and Henry Jenkins. "Confronting Horror, Embracing Fantasy: A Conversation about Lovecraft Country and Radical Imagination in Higher Education." *Policy Futures in Education* 20, no. 1 (2022): 73–85.

Hatrick, Jessica, and González, Olivia. "*Watchmen*, Copaganda, and Abolition Futurities in US Television." *Lateral: Journal of The Cultural Studies Association* 11, no. 2 (Fall 2022). https://csalateral.org/issue/11-2/watchmen-copaganda-abolition-futurities-us-television-hatrick-gonzalez/.

Hauskeller, Michael, Thomas D. Pillbeck, and Curtis D. Carbonell, eds. *The Palgrave Handbook of Posthumanism in Film and Television*. New York: Palgrave Macmillan, 2015.

Hawkes, Lisa D. "Hippolyta's Spiritual Awakening Through Spiritual Warfare in *Lovecraft Country* (2020)." *First Thoughts on Lovecraft Country*, a special issue of *Studies in the Fantastic* 12 (Winter 2021): 1–17.

Haylock, Zoe. "*Being Mary Jane* Season 4, Episode 15 Recap: Feeling Hashtagged." *Refinery*, August 15, 2017. www.refinery29.com/en-us/2017/08/168147/being-mary-jane-recap-season-4-episode-15.

Heitner, Devorah. *Black Power TV*. Durham, NC: Duke University Press, 2013.

Henderson, Felicia D. "From Sitcom Girl to Drama Queen: Soul Food's Showrunner Examines Her Role in Creating TV's First Successful Black-Themed Drama." In *Watching While Black Rebooted!: The Television and Digitality of Black Audiences*, ed. Beretta E. Smith-Shomade, 57–71. New Brunswick, NJ: Rutgers University Press, 2023.

Hensley, Ellie. "Why Atlanta's warehouse owners are jumping into the movie business." *The Business Journals*, November 1, 2014. www.bizjournals.com/bizjournals/news/2014/11/01/why-atlantas-warehouse-owners-are-jumping-into-the.html.

Hermann, Isabella. "Artificial Intelligence in Fiction: Between Narratives and Metaphors." *AI and Society* 38 (2023): 319–329.

Historic Oakland Foundation. "Character Areas and Landmarks." Oakland Cemetery, Atlanta. https://oaklandcemetery.com/character-areas-and-landmarks/.

Ho, Randy. "Black Residents of Trilith Sue Studio, Development for Discrimination." *Atlanta Journal-Constitution*, November 17, 2022. www.ajc.com/life/radiotvtalk-blog/black-residents-of-trilith-sue-studio-development-for-discrimination/4ALI4D33UFFQFLUXQTHN5M3T64/.

Hoberek, Andrew. "Of Watchmen and Great Men: The Graphic Novel, The Television Series, and the Police." *Literature/Film Quarterly* 49, no. 4 (Fall 2021). https://lfq.salisbury.edu/_issues/49_4/of_watchment_and_great_men_the_graphic_novel_the_television_series_and_the_police.html.

Hobson, Maurice J. *The Legend of the Black Mecca: Politics and Class in the Making of Modern Atlanta*. Chapel Hill: University of North Carolina Press, 2017.

Hollis, Henry. "Tyler Perry Donates $50K in Gift Cards to Peoplestown Community Through APD." *Atlanta Journal-Constitution*, July 17, 2020. www.ajc.com/news/tyler-perry-donates-50k-in-gift-cards-to-peoplestown-community-through-apd/Z4IV5ZOZ4FD4RKBGMU5OZ6ZC5I/.

Honeycutt, John. "Albuquerque Reaches $42.5K Settlement in Excessive Force Case." *KRQE News*, September 1, 2023. www.krqe.com/news/albuquerque-metro/albuquerque-reaches-42-5k-settlement-in-excessive-force-case/.
Hong, Catherine. "Step Onto Tyler Perry's 300-Acre Production Studio." *Architectural Digest*, November 7, 2019. www.architecturaldigest.com/story/step-onto-tyler-perrys-300-acre-production-studio.
Immergluck, Dan. *Red Hot City: Housing, Race, and Exclusion in Twenty-First Century Atlanta*. Oakland: University of California Press, 2022.
Jacobson, Brian R., ed. *In the Studio: Visual Creation and its Material Environments*. Oakland: University of California Press, 2020.
Jakle, Jeanne. "Longoria Lauds 'Devious Maids' Amid Heated Debate." *My San Antonio*, June 25, 2013. www.mysanantonio.com/entertainment/entertainment_columnists/jeanne_jakle/article/Longoria-lauds-Devious-Maids-amid-heated-debate-4613193.php.
Jenkins, Destin, and Justin Leroy, eds. *Histories of Racial Capitalism*. New York: Columbia University Press, 2021.
Jenner, Mareike. *Netflix and the Reinvention of Television*. New York: Palgrave Macmillan, 2018.
Jimenez, Carlos, and Alfredo Huante. "Home in Vida and Gentefied: The Politics of Representation in Gente-fication Narratives." *Aztlan: A Journal of Chicano Studies* 48, no. 1 (Spring 2023): 21–51.
Jones, Barron, and Lalita Moskowitz. "The Inhumane Conditions at MDC are a Result of Over-Incarceration," *ACLU New Mexico*, July 15, 2022. www.aclu-nm.org/en/news/inhumane-conditions-mdc-are-result-over-incarceration.
Joseph, Ralina L. *Postracial Resistance: Black Women, Media, and the Uses of Strategic Ambiguity*. New York: New York University Press, 2018.
Joyner, Chris. "Georgia Capitol Heavy with Confederate Symbols." *Atlanta Journal-Constitution*, September 4, 2015. www.ajc.com/news/state--regional-govt--politics/georgia-capitol-heavy-with-confederate-symbols/z051suE0a7bqO5cWhXlZnJ/.
Joyrich, Lynne. "American Dreams and Demons: Television's 'Hollow' Histories and Fantasies of Race." *The Black Scholar* 48, no. 1 (2018): 31–42.
Kaplan, Elsie. "APD: Former Spokesman Simon Drobik was 'Gaming the System.'" *Albuquerque Journal*, October 23, 2020. www.abqjournal.com/news/local/apd-former-spokesman-simon-drobik-was-gaming-the-system/article_76a55a69-4937-5ea2-8223-1e6d7f5100e1.html.
———. "APD Officer Violated Policies in Inmate's Suicide." *Albuquerque Journal*, October 6, 2020. www.abqjournal.com/news/local/apd-officer-violated-policies-in-inmates-suicide/article_73d18188-e036-5597-854f-238003a68089.html.
Kelley, Robin D. G. "Birth of a Nation Redux: Surveying Trumpland with Cedric Robinson." *Boston Review*, November 5, 2020. www.bostonreview.net/articles/robin-d-g-kelley-births-nation/.
Kelley, Sonaiya. "Nichole Beharie was Called 'Problematic' and Blacklisted. 'Miss Juneteenth' Brings Redemption." *San Diego Union-Tribune*, June 19, 2020. www

.sandiegouniontribune.com/entertainment/movies/story/2020-06-19/nicole-beharie-miss-juneteenth.

Kent, John. "Culver City: From Whites Only to National Model of Diversity and Inclusion?" (slideshow). n.d. www.culvercity.org/files/assets/public/v/2/documents/planning-amp-development/advance-planning/speaker-series/191121_discriminatory-land-use-policies/speakerseriesdiscriminatoryslides.pdf.

Kilkenny, Katie. "Tyler Perry Puts $800M Studio Expansion on Hold After Seeing OpenAI's Sora: 'Jobs Are Going to Be Lost.'" *Hollywood Reporter*, February 22, 2024. www.hollywoodreporter.com/business/business-news/tyler-perry-ai-alarm-1235833276/.

Klein, Gary. "Mill Valley Native Sues TV Show in Copyright Case." *Marin Independent Journal*, October 7, 2006. www.marinij.com/2006/10/07/mill-valley-native-sues-tv-show-in-copyright-case/.

———. "Tam High Grad, TV Networks Settle Suit." *Marin Independent Journal*, September 3, 2007. www.marinij.com/2007/09/03/tam-high-grad-tv-networks-settle-suit/.

Koshy, Susan, Lisa Marie Cacho, Jodi A. Byrd, and Brian Jordan Jefferson, eds. *Colonial Racial Capitalism*. Durham, NC: Duke University Press, 2022.

Krasnow, Bruce, and Dan Schwartz. "Santa Fe Studios Clashes With County Over Debt Repayment, Expansion." *Santa Fe New Mexican*, December 13, 2015. www.santafenewmexican.com/news/local_news/santa-fe-studios-clashes-with-county-over-debt-repayment-expansion/article_f6e830fa-a308-5e6b-8285-d3d4fdb18a41.html.

Lagerwey, Jorie, and Taylor Nygaard. *Horrible White People: Gender, Genre, and Television's Precarious Whiteness*. New York: New York University Press, 2020.

Lambe, Stacey. "How 'Antebellum,' 'Lovecraft County' and More Horror Stories are Centering Black Experiences." *Entertainment Weekly*, September 25, 2020. www.etonline.com/how-antebellum-lovecraft-country-and-more-horror-stories-are-centering-black-experiences-153705.

———. "'Lovecraft Country' Creator Says There are 'Seasons Upon Seasons' of Stories Left to Tell." *Entertainment Tonight*, October 18, 2020. www.etonline.com/lovecraft-country-creator-on-future-seasons-and-changes-from-book-154873.

Leishman, Rachael. "Tamara Podemski Brings Change to the Western Genre as Joy in 'Outer Range.'" *The Mary Jane*, May 13, 2022. www.themarysue.com/interview-tamara-podemski-outer-range/.

Lewis, Joseph. "Man's Fear of a Black Planet: Monstrous Ontological Encounters at Sundown in HBO's *Lovecraft Country*." *Studies in the Fantastic* 12 (Winter 2021): 18–37.

Ligozat, Ann-Laure, Julien Lefèvre, Aurélie Bugeau, and Jacques Combaz. "Unraveling the Hidden Environmental Impacts of AI Solutions for Environment Life Cycle Assessment of AI Solutions." *arXiv*, April 12, 2022. https://arxiv.org/abs/2110.11822v2.

Lipsitz, George. *How Racism Takes Place*. Philadelphia: Temple University Press, 2011.

Maddaus, Gene. "Georgia Film Credit Grows to Record $1.3 Billion." *Variety*, January 17, 2023. https://variety.com/2023/tv/news/georgia-film-credit-hits-record-1235488240/.
Manigault-Bryant, LeRhonda S., Tamura A. Lomax, and Carol B. Duncan, eds. *Womanist and Black Feminist Responses to Tyler Perry's Productions*. New York: Palgrave Macmillan, 2014.
Marez, Curtis. *Drug Wars: The Political Economy of Narcotics*. Minneapolis: University of Minnesota Press, 2004.
———. "'Plenty of Room to Swing a Rope': *Watchmen* and the Racial Politics of Place." In *After Midnight: Watchmen After Watchmen*, ed. Drew Morton, 143–154. Jackson: University Press of Mississippi, 2022.
———. *University Babylon: Film and Race Politics on Campus*. Oakland: University of California Press, 2019.
Marini, Anna Marta. "*Gentefied* and the Representation of Gentrification Related Latinx Conflicts." *JAm It!* 4 (May 2021): 35–57.
Maxouris, Christina. "Atlanta Wants to Build a Massive Police Training Facility in a Forest. Neighbors are Fighting to Stop It." *CNN*, September 24, 2022. www.cnn.com/2022/09/24/us/atlanta-public-safety-training-center-plans-community/index.html.
Mayer, Vicki. *Almost Hollywood, Nearly New Orleans: The Lure of the Local Film Economy*. Oakland: University of California Press, 2017.
———. *Below the Line: Producers and Production Studies in the New Television Economy*. Durham, NC: Duke University Press, 2011.
———, ed. *Special Issue on Treme. Television and New Media* 13, no. 4 (May 2012).
Mayer, Vicki, Miranda J. Banks, and John T. Caldwell, eds. *Production Studies: Cultural Studies of Media Industries*. New York: Routledge, 2009.
McNear, Claire. "Ranking the Cameos From 'Breaking Bad' and 'Better Call Saul' in 'El Camino.'" *The Ringer*, October 15, 2019. www.theringer.com/movies/2019/10/15/20914631/el-camino-netflix-cameos-ranked-breaking-bad-better-call-saul.
McNutt, Myles. *Television's Spatial Capital: Location, Relocation, Dislocation*. New York: Routledge Press, 2022.
Melamed, Jodi. "Diversity." In *Keywords for American Cultural Studies*, eds. Bruce Burgett and Glenn Hendler, 93–97. New York: New York University Press, 2020.
———. "Racial Capitalism." *Critical Ethnic Studies* 1, no. 1 (Spring 2015): 76–85.
Miller, Toby, Nitin Govil, John McMurria, Richard Maxwell, and Ting Wang. *Global Hollywood 2*. 2nd Edition. London: British Film Institute, 2004.
Mitchell, Nick. "Diversity." In *Keywords for African American Studies*, ed. Erica R. Edwards, Roderick A. Ferguson, and Jeffrey O. G. Ogbar, 69–74. New York: New York University Press, 2018.
Molina-Guzmán, Isabel. *Latinas and Latinos on TV: Colorblind Comedy in the Postracial Network Era*. Tucson: University of Arizona Press, 2018.
Moore, Jazmyn T., Jessica N. Ricaldi, Charles E. Rose, et al. "Disparities in Incidence of Covid-19 Among Underrepresented Racial/Ethnic Groups in Counties Identified as Hotspots During June 5–18, 2020–22 States, February-June 2020." Centers for

Disease Control and Prevention, August 12, 2020. www.cdc.gov/mmwr/volumes/69/wr/mm6933e1.htm#:~:text=During%20June%205–18%2C%20205,more%20underrepresented%20racial%2Fethnic%20groups.

Myrick, Kevin. "No Plans to Remove Polk County's Monument for Confederate Soldiers." *The Polk County Standard Journal*, August 17, 2017. www.northwestgeorgianews.com/polk_standard_journal/news/local/no-plans-to-remove-polk-county-s-monument-for-confederate/article_7d822454-8361-11e7-b636-0f54981c180e.html.

New Mexico Corrections Department. "Old Main/Filming and Tours." December 8, 2024, www.cd.nm.gov/divisions/corrections-industries/old-main/old-main-filming-tours/.

New Mexico Indian Affairs Department. "Nations, Pueblos, and Tribes." December 8, 2024, www.iad.nm.gov/nations-pueblos-and-tribes/.

Newsom, Bree. "Go Ahead, Topple the Monuments to the Confederacy, All of Them." *Washington Post*, August 18, 2017. www.washingtonpost.com/opinions/go-ahead-topple-the-monuments-to-the-confederacy-all-of-them/2017/08/18/6b54c658-8427-11e7-ab27-1a21a8e006ab_story.html.

Nguyen, Viet Thanh. "How 'Watchmen's' Misunderstanding of Vietnam Undercuts its Vision of Racism." *Washington Post*, December 18, 2019. www.washingtonpost.com/outlook/2019/12/18/how-watchmens-misunderstanding-vietnam-undercuts-its-vision-racism/.

Noble, Safiya Umoja. *Algorithms of Oppression: How Search Engines Reinforce Racism*. New York: New York University Press, 2018.

Noriega, Chon. *Shot in America: Television, the State, and the Rise of Chicano Cinema*. Minneapolis: University of Minnesota Press, 2000.

O'Brian, Kean, Leonardo Vilchis, and Corina Maritescu. "Boyle Heights and the Fight against Gentrification as State Violence." *American Quarterly* 71, no. 2 (June 2019): 389–386.

O'Connell, Mickey. "'Reservation Dogs' Boss on Combatting Indigenous Stereotypes, Embracing Gripes and That Emmy Snub." *Hollywood Reporter*, August 18, 2022. www.hollywoodreporter.com/tv/tv-features/reservation-dogs-sterlin-harjo-emmy-snub-indigenous-stereotypes-1235199618/.

O'Keefe, Meghan. "HBO's 'Watchmen' was Ahead of its Time—By 9 Months" *Decider*. June 4, 2020. https://decider.com/2020/06/04/watchmen-on-hbo-2020-relevance-tulsa-massacres/.

O'Mahony, Lauren, Melissa Merchant, and Simon Order. "NecroPolitics in a Post-Apocalyptic Zombie Diaspora: The Case of AMC's *The Walking Dead*." *Journal of Postcolonial Writing* 57, no. 1 (2021): 89–103.

Orenstein, Dana. *Out of Stock*. Chicago: University of Chicago Press, 2019.

Page, Allison, and Laurie Ouellette. "The Prison-Televisual Complex." *International Journal of Cultural Studies* 23, no. 1 (2019): 1–17.

Palmiera, Lea. "'Get Shorty' Season 3." *Decider*. October 4, 2019. https://decider.com/2019/10/04/chris-odowd-interview-get-shorty-epix/.

Parks, Lisa. "Falling Apart: Electronics Salvaging and the Global Media Economy." In *Residual Media*, ed. Charles Acland, 32–47. Minneapolis: University of Minnesota Press, 2007.

Parmett, Helen Morgan. *Down in Treme: Race, Place, and New Orleans on Television*. Stuttgart: Franz Steiner Verlag Wiesbaden GmbH, 2019.

———. "Site-Specific Television as Urban Renewal: Or, How Portland became *Portlandia*." *International Journal of Cultural Studies* 21, no. 1 (2018): 42–56.

Penney, Joe. "How Medical Negligence at the US Border Killed an Immigrant Father." *The Nation*, February 25, 2020. www.thenation.com/article/world/ice-death-negligence/.

Perkins, Nichole. "*Greenleaf* Recap: Black and Blue." *Vulture*, June 29, 2016. www.vulture.com/2016/06/greenleaf-recap-season-1-episode-4.html.

Perry, Tyler. "Tyler Perry Gives Exclusive Interview On *Joe Scarborough Presents*." Tyler Perry.com. https://tylerperry.com/tyler-perry-gives-exclusive-interview-on-joe-scarborough-presents/.

———. "Tyler Perry Studios Welcomed the Atlanta Police Department in Appreciation Luncheon." Tyler Perry.com. https://tylerperry.com/tyler-perry-studios-welcomed-the-atlanta-police-department-in-appreciation-luncheon/.

———. "Tyler Perry's Oscar Speech Delivers a Message of Hope and a Refusal of Hate." Tyler Perry.com. https://tylerperry.com/tyler-perrys-oscar-speech-delivers-a-message-of-hope-and-a-refusal-of-hate/.

Petski, Denise. "Nichole Beharie Speaks Out About Her 'Sleepy Hollow' Exit and the Controversy's Impact on her Career." *Deadline*, June 20, 2020. https://deadline.com/2020/06/nicole-beharie-sleepy-hollow-exit-controversy-impact-career-1202965074/.

Phillips, Jay. "Park Closure Sparks Talk of Ongoing Lawsuit Against DeKalb and Blackhall." *The Champion*, July 25, 2020. https://thechampionnewspaper.com/park-closure-sparks-talk-of-ongoing-lawsuit-against-dekalb-and-blackhall/.

Pinto, Nick. "When Cops Break Bad: Inside a Police Force Gone Wild." *Rolling Stone*, January 29, 2015, 5–6. www.rollingstone.com/culture/culture-news/when-cops-break-bad-inside-a-police-force-gone-wild-69487/.

Prison Policy Initiative. "Georgia Profile." December 9, 2024. www.prisonpolicy.org/profiles/GA.html.

Private-Jet.pro. "The Most Expensive Private Jet in the World." https://private-jets.pro/en/guide/most-expensive-private-jet-world.html.

Pulido, Laura, Laura Barraclough, and Wendy Cheng. *A People's Guide to Los Angeles*. Berkeley: University of California Press, 2012.

Ramón, Ana-Christina, Michael Tran, and Darnell Hunt. *Hollywood Diversity Report 2023, Part 2: Television*. Los Angeles: UCLA Entertainment and Media Research Initiative, 2023.

Rappas, Ipek A. Celik. "From *Titanic* to *Game of Thrones*: Promoting Belfast as a Global Media Capital." *Media, Culture, and Society* 41, no. 1 (January 2019): 1–18.

Regester, Charlene B. "African American Extras in Hollywood During the 1920s and 1930s." *Film History* 9 (1997): 95–115.

Reiff-Shanks, Mychal. "The Utilization of Historical Reenactment in *Lovecraft Country*." *Studies in the Fantastic* 12 (Winter 2021): 55–73.

Reyes-Velarde, Alejandra. "Drama on Set and Off; 'Vida,' a TV Show About Gentrification, Faces Protests by Some Who Say it is Doing Just That." *Los Angeles Times*, September 1, 2018, A.1.

Ristau, Reece. "Eva Longoria Talks 'Devious Maids' Backlash, Latino Perceptions at Produced By Conference." *Variety*, May 31, 2015. https://variety.com/2015/tv/news/eva-longoria-devious-maids-latino-1201509025/.

Rivero, Yeidy. *Tuning Out Blackness: Race and Nation in the History of Puerto Rican Television*. Durham, NC: Duke University Press, 2005.

Robbins, Scott, and Aimee van Wynsberghe. "Our New Artificial Intelligence Infrastructure: Becoming Locked into an Unsustainable Future." *Sustainability* 14 (2024): 1–11. www.mdpi.com/2071-1050/14/8/4829.

Robinson, Cedric J. *Black Marxism: The Making of the Black Radical Tradition*. Chapel Hill: University of North Carolina Press, 1983.

———. *Forgeries of Memory and Meaning: Blacks and the Regimes of Race in American Theater and Film Before World War II*. Chapel Hill: University of North Carolina Press, 2007.

Rodríguez, Dylan. *White Reconstruction: Domestic Warfare and the Logics of Genocide*. New York: Fordham University Press, 2020.

Romero, Ariana. "The True Story of *Teenage Bounty Hunters*, Straight From Its Creator." *Refinery 29*, August 14, 2020. https://ew.com/tv/teenage-bounty-hunters-netflix-kathleen-jordan-preview/; https://www.refinery29.com/en-us/2020/08/9966810/is-teenage-bounty-hunters-true-story-netflix.

Rossouw, Martin P. "*Watchmen*: From Co-Mix to Remix." A special issue of *Literature/Film Quarterly* 49, no. 4 (Fall 2021).

Rubin, Jennifer. "How the Republican Party Became a Death Cult." *Washington Post*, June 23, 2020. www.washingtonpost.com/opinions/2020/06/23/how-republican-party-became-death-cult/.

Russworm, TreaAndrea M., Samantha N. Sheppard, and Karen M. Bowdre, eds. *From Madea to Media Mogul: Theorizing Tyler Perry*. Jackson: University Press of Mississippi, 2016.

Ryan, Maureen. *Burn It Down: Power, Complicity, and a Call for Change in Hollywood*. New York: Mariner Books, 2003.

Sacks, Briana. "Trilith Promised a New Model for Hollywood in the Atlanta Suburbs, But Black Residents and Former Employees Say Racism Lies Just Below the Surface." *BuzzFeed News*, July 18, 2022. www.buzzfeednews.com/article/briannasacks/trilith-georgia-studios-town-race.

Saha, Anamik. "Production Studies of Race and the Political Economy of Media." *Journal of Cinema and Media Studies* 60, no. 1 (Fall 2020): 138–142.

Sánchez, George J. "Why Are Multiracial Communities So Dangerous?: A Comparative Look at Hawai'i; Cape Town, South Africa; and Boyle Heights, California." *Pacific Historical Review* 86, no. 1 (2017): 153–170.

Sánchez, Rosaura, and Beatrice Pita. *Spatial and Discursive Violence in the U.S. Southwest*. Durham, NC: Duke University Press, 2021.

Sanson, Kevin. *Mobile Hollywood: Labor and the Geography of Production*. Oakland: University of California Press, 2024.

Saperstein, Pat. "Is Trilith, a 'Company Town' for Marvel's Georgia Production Workers, the Template of the Future?" *Variety*, March 10, 2022. https://variety.com/2022/film/news/marvel-trilith-town-georgia-1235194027/.

Saporta, Maria. "Fort Mac-Tyler Perry Deal Could Close Friday; Sen. Vincent Fort Declares Sale Illegal." *Atlanta Business Chronicle*, June 26, 2015. www.bizjournals.com/atlanta/morning_call/2015/06/fort-mac-tyler-perry-deal-could-close-friday-sen.html.

Sassaki, Raphael. "Moore on Jerusalem, Eternalism, Anarchy, and Herbie!" *Alan Moore World*, November 18, 2019.

Savage, Kirk. *Standing Soldiers, Kneeling Slaves: Race, War, and Monument in Nineteenth-Century America*. Princeton, NJ: Princeton University Press, 1999.

Schager, Nick. "The Gloriously Bizarre Genius of Benny Safdie." *Daily Beast*, December 4, 2023. www.thedailybeast.com/obsessed/benny-safdie-interview-on-the-bizarre-genius-of-the-curse.

Schmieding, Jana (@janaunplgd). "Hope you enjoyed the show, babes!" Instagram, April 16, 2021. www.instagram.com/p/CNv_cCCLvLK/.

Schulman, Sandra Hale. "Art Installations Remind LA Residents They are on 'TONGVALAND.'" *Indian Country Today*, August 30, 2021. https://ictnews.org/news/art-installations-remind-la-residents-theyre-on-tongvaland.

Schwartz, Stephanie A. "Consciousness, Covid, and the Rise of an American Death Cult." *Explore* 18, no. 3 (May–June 2022): 259–263. www.ncbi.nlm.nih.gov/pmc/articles/PMC8935966/.

Sepinwall, Alan. "'There's Power in Laughing at the Pain': 'Brockmire' Creator on Series' American Nightmare." *Rolling Stone*, May 5, 2020. www.rollingstone.com/tv-movies/tv-movie-features/brockmire-finale-creator-church-cooper-interview-990616/.

Shannon, Jerry, Amanda Abraham, Grace Bagwell Adams, and Matthew Hauer. "Racial Disparities for Covid 19 Mortality in Georgia: Spatial Analysis by Age Based on Excess Deaths." *Social Science and Medicine*, January 2022. www.ncbi.nlm.nih.gov/pmc/articles/PMC8734109/#:~:text=Conclusions,to%20social%20determinants%20of%20health.

Siebler, Kay. "Black Feminists in Serialized Dramas: The Gender/Sex/Race Politics of *Being Mary Jane* and *Scandal*." *Journal of Popular Film and Television* 46, no. 3 (September 10, 2019): 152–62.

Simek, Nicole. "Speculative Futures: Race in *Watchmen*'s Worlds." *Symploke* 28, nos. 1–2 (2020): 385–404.

Sisson, Patrick. "Albuquerque is Winning the Streaming Wars." *Bloomberg*, May 3, 2021. www.bloomberg.com/news/articles/2021-05-03/why-hollywood-is-moving-to-albuquerque.

Smith-Shomade, Beretta E. *Pimpin' Ain't Easy: Selling Black Entertainment Television.* New York: Taylor and Francis, 2008.

———, ed. *Watching While Black: Centering the Television of Black Audiences.* New Brunswick, NJ: Rutgers University Press, 2012.

———, ed. *Watching While Black Rebooted!: The Television and Digitality of Black Audiences.* New Brunswick, NJ: Rutgers University Press, 2023.

Southern Poverty Law Center. "Flags and Other Symbols used by Far-Right Groups in Charlottesville." Southern Poverty Law Center. August 12, 2017. www.splcenter.org/hatewatch/2017/08/12/flags-and-other-symbols-used-far-right-groups-charlottesville

Spangler, Todd. "Netflix is Paying Less Than $30 Million for Albuquerque Studios, Which Cost $91 Million to Build." *Variety*, October 16, 2018. https://variety.com/2018/digital/news/netflix-albuquerque-studios-deal-terms-30-million-1202981274/.

Spry, Jeff. "*Outer Range* Star Tamara Podemski Explains her Cliffhanger Ending and Season 2 Hopes." *Inverse*, May 13, 2022. www.inverse.com/entertainment/outer-range-tamara-podemski-cliffhanger-season-2.

Steinhart, Daniel. *Runaway Hollywood: Internationalizing Postwar Production and Location Shooting.* Oakland: University of California Press, 2019.

Strykowski, Jason. *A Guide to New Mexico Film Locations: From Billy the Kid to Breaking Bad and Beyond.* Albuquerque: University of New Mexico Press, 2021.

Sugg, Katherine. "*The Walking Dead*: Late Liberalism and Masculine Subjection in Apocalypse Fictions." *Journal of American Studies* 49, no. 4 (2015): 793–811.

Tahmahkera, Dustin. *Tribal Television: Viewing Native People in Sitcoms.* Chapel Hill: University of North Carolina Press, 2014.

Tatum, Gloria. "Native Americans Share Concerns Over Fate of Forest." *Streets of Atlanta*, May 2, 2022. https://streetsofatlanta.blog/2022/05/02/native-americans-share-concerns-over-fate-of-forest/.

Third Rail Studios. "Third Rail Studios Info Sheet." December 9, 2024. https://thirdrailstudios.com/wp-content/uploads/Third-Rail-Studios-Info-Sheet-20190901.pdf.

Thomas, Tony. "Local Jail Reaps Benefits of Georgia's Booming Film Industry." *WSB-TV*, November 26, 2016. www.youtube.com/watch?v=oAi9MpbwDpU.

Torres, Sasha. *Black, White, and in Color: Television and Black Civil Rights.* Princeton, NJ: Princeton University Press, 2003.

Tran, Diep. "'The Cleaning Lady' Aims to Break Stereotypes About Asian Workers in the US." *NBC News*, January 31, 2022. www.nbcnews.com/news/asian-america/-cleaning-lady-aims-break-stereotypes-asian-workers-us-rcna14270#:~:text ="There%20is%20actually%20a%20greater,to%20center%20a%20Filipino%20 family.

Travers, Ben. "Nathan Fielder and Benny Safdie Nearly Talked Themselves Out of Making 'The Curse.'" *Indiewire*, December 4, 2023. www.indiewire.com/features/interviews/nathan-fielder-benny-safdie-the-curse-interview-1234931493/.

Turner, Jacob S. and Lisa G. Perk. "White Men Holding on for Dear Life and Taking It: A Content Analysis of the Gender and Race of the Victims and Killers in *The Walking Dead*." *Sex Roles* 81 (2019): 660.

Ultra-red. "*Desarmando Desarrollismo*: Listening to Anti-Gentrification in Boyle Heights." *Field: A Journal of Socially Engaged Art Criticism*, Fall, 2019. http://field-journal.com/issue-14/desarmando-desarrollismo.

Upton, Dell. *What Can and Can't be Said: Race, Uplift, and Monument Building in the Contemporary South*. New Haven, CT: Yale University Press, 2015.

Villarreal, Yvonne. "Netflix Tackles Gentrification; 'Gentefied' Debuts Amid Concern It May Contribute to Problem in Boyle Heights." *Los Angeles Times*, February 2020, E1.

Vives, Ruben. "As Rents Soar in L.A., Even Boyle Heights' Mariachis Sing the Blues." *Los Angeles Times*, September 9, 2017. www.latimes.com/local/lanow/la-me-boyle-heights-musicians-gentrification-20170909-htmlstory.html.

Wakefield, Stephanie, and Glenn Dyer. "Stop the Metaverse, Save the Real World." *e-flux*, September, 2022. www.e-flux.com/architecture/horizons/493130/stop-the-metaverse-save-the-real-world/.

Warner, Kristen J. "Being Mary Jane: Cultural Specificity." In *How to Watch TV*, ed. Ethan Thompson and Jason Mittell, 109–16. New York: New York University Press, 2020.

———. *The Cultural Politics of Colorblind TV Casting*. New York: Routledge, 2015.

West Georgia Textile Heritage Trail. "Cedartown." December 9, 2024. https://westgatextiletrail.com/cedartown/.

Whitehead, Nadia. "What Was it Like to Consult for Breaking Bad?" *Science*, May 2, 2014. www.science.org/content/article/what-was-it-consult-breaking-bad.

Whitted, Qiana. *EC Comics: Race, Shock, and Social Protest*. New Brunswick, NJ: Rutgers University Press, 2019.

Whyte, Gabrielle. "Seeing Color: The Relationship Between Popular Media and Anti-Racist Social Justice." *The Macksey Journal* 4, no. 57 (2023): 1–11.

Wiggins, David N. *Georgia's Confederate Monuments and Cemeteries*. New York: Arcadia Publishing, 2006.

Wilson, Chris. "The Historic Railroad Buildings of Albuquerque: An Assessment of Significance." Wheels Museum, Albuquerque, NM, December 9, 2024. https://wheelsmuseum.org/wp-content/uploads/2015/11/The-Historic-Railroad-Buildings-of-Albuquerque.pdf1.

Wood, Robin. "An Introduction to American Horror." In *Robin Wood on the Horror Film: Collected Essays and Reviews*, ed. Barry Keith Grant, 114–67. Detroit: Wayne State University Press, 2018.

Woods, Clyde. "Les Misérables of New Orleans: Trap Economics and the Asset Stripping Blues, Part 1." *American Quarterly* 61, no. 3 (September 2009): 769–96.

Wooten, Jillian. "An Historical Analysis of The Atlanta Prison Farm." Atlanta City Planning Office. November 5, 1999. https://dekalbhistory.org/wp-content/uploads

/2019/11/historical-analysis-of-honor-farm.pdf; https://www.ajc.com/blog/buzz/old-atlanta-prison-farm-favorite-with-filmmakers/yHzdSgh4ywGlzFaxjokqdL/.

Wu, Sky. "It Takes a Character to Make a Character—Alumni Profile, Leah Longoria '12." *The Amherst Student*, November 13, 2021. https://amherststudent.com/article/it-takes-a-character-to-make-a-character/.

Zahirovic-Herbert, Velma, and Karen M. Gibler. "The Effect of Film Production Studios on Housing Prices in Atlanta, the Hollywood of the South." *Urban Studies* 59, no. 4 (July 29, 2021): 771–788.

INDEX

Akil, Mara Brock, 20, 121, 135–41
Akil, Salim, 20, 141
Albuquerque, New Mexico, 2, 6–7, 40–42, 48–51; Central Avenue, 85–86; Journal Studio Center, 44; Netflix Albuquerque Studios, 11, 33, 36–37, 42, 49, 51, 56, 59, 84, 147, 151, 168; Sundowner Motel, 85–86, *86*
Albuquerque Municipal Detention Center, 59, 62, *62*
Albuquerque Police Department, 53, 53–55, 151
Albuquerque Rail Yards, 31–34, *33*, 38, 40–41, 84
Alexander-Floyd, Nikol G., 126, 141–42
Alvitre, Weshoyot, 171–72
Ambitions, 20, 23, 26, 119
Anthropocene, 46–47
anti-gentrification protests, 159–67, *162*
architecture: Pueblo-style, 17, 36, 75; symbolic meaning of, 16–17, 74–75, 80, 85–86, 137, 140, 144–46, 155; "territorial revival style" of, 17, 74–75
artificial intelligence, 177–80
Atlanta, 19–20, 73, 76–79, 82, 93, 127
Atlanta, Georgia, 5, 6–7, 11, 30–31, 34–38, 43–44, 47–48, 51, 56–57, 61, 64–65, 69, 78–79, 87, 96, 98–99, 105, 107, 118–43, 144, 149–50, 151. *See also* Shadowbox Studios
Atlanta Police Department, 132, 138
The Avengers, 33, 78

Back to the Future, 171
Bakersfield, California, 100, 164, 167

Banks, Jonathan ("Mike Ehrmantraut"), 31, 46, *53*
Beharie, Nicole, 88–89
Being Mary Jane, 7, 19–20, 119–23, 136–39
bell hooks, 122
Better Call Saul, 31, 43, 53, *53*, 57, 59–60, 60, 147
Big Sky, 59, 63, 64
Black creatives, 18–21, 23, 78; privileged, 80–82, 91, 119, 142
Blackhall Studios, 37, 47–48, 56, 59, 92, 156, 168, 173. *See also* Shadowbox Studios
Black Lightening, 20, 96, 98, 107–8, *108*
Black Marxism (Robinson), 8
Black Panther, 78, 129, 149
Black racial capitalism in place, 22–24, 80–82, 91, 119, 121–22, 126
Bosch, 153–54
Boyle Heights, 151–54, 159–67
Boyle Heights Against Artwashing and Displacement (BHAAD), 153, 161
Breaking Bad, 4, 7, 31, 43, 45–46, 53–57, 59, 62, 147; To'hajiilee Navajo settlement, 49–51, *50*
Briarpatch, 31–32, 41, 54
Brockmire, 177–78
Broderick, Matthew, 40
Butler, Octavia, 122

cannibalism, 40, 48
Carter, Jimmy, 4
cattle, 40–41, 86, 113
Cedartown, Georgia, 100–101
Central Avenue, Albuquerque, 85–86

229

Charlottesville, Virgina, 107, 123, 173
Chattahoochee River, 37, 77–78, 150
Chávez, Linda Yvette, 167
The Cleaning Lady, 32, 44–45, *45*, 54, 68–71, *70*, *71*
colorblind casting, 18–19, 136
Confederate flag, 65, 83–84, 97, 123, 170
Confederate memorials, 65, 78, 95–98, 100–103, 105–8, *107*, *108*, 170
copaganda, 61–62
"Cop City" (Georgia Department of Corrections), 11, 36, *36*, 47, 56–57, 92, 132
COVID-19, 72
Cranston, Bryan ("Walter White"), 31, 43, 45–46, 50
Creepshow, 144–45, 149
Culver City, California, 9
The Curse, 17, 155–58
Custer, George Armstong, 48

David, Keith, 118, 144
Daybreak, 32, 40–41
Decatur, Georgia, 101, 106
DEI horror: serial killer, 73–82; supernatural, 82–92
DEI police, 62–68
DEI TV, 17–19, 22–24, 46–47, 93–94, 111, 113–17, 136, 143, 149
Department of Homeland Security, 56
The Deputy, 51, 63–64, *64*, 66
Devious Maids, 6–7, *7*, 18–19, 43–44, 68–71
Diamond Handcuffs, 9
Doom Patrol, 95
Drug Enforcement Agency, 11, 49–50, 56, 65
The Dukes of Hazzard, 65

Eagle Rock Studios, Gwinnett County, Georgia, 30, 59, 69
EC Comics, 144
Echo, 52, 149
Electric Owl Studios, 35
Española, New Mexico, 156–58

Estancia, New Mexico, 87
EUA/Screen Gems Studios, 69

Falling for Angels, 159–60
Fanon, Frantz, 122
Farmer, Gary, 114, 167
Faye, Jennifer, 46–47
Fayetteville, Georgia, 147–48, *148*
film and TV production studies, 16
film incentives, 2–6, 14–15, 55, 69, 87, 104–5, 113, 128, 139, 181
Floyd, George, 22, 81–82, 99, 125, 132
forests, 25, 36–38, 47, 57, 76, 87, 90, 92, 116, 150, 152, 173
Forgeries of Memory and Meaning (Robinson), 8, 10, 22–23
foster care, 64, 73–76, 78–80, 82, 93, 108, 110
Foucault, Michel, 159

Gang Related, 153
Garner, Eric, 124–25
Gentefied, 145–46, 163–67
Georgia Bureau of Investigation, 64
Get Shorty, *1*, 1–2, *2*, 28
Gilligan, Vince, 46
The Girlfriend Experience, 7, 136
Glover, Donald, 78–79, 82
Goliath, 153–54
Good Girls, 51
Gray, Herman, 12, 124–25
Green, Misha, 20, 90
Greenleaf, 60, 118–20, *119*, *120*, 123, 132–33, 142–43, *143*
Greenwood Massacre, 98–100, 103–4
Gremlins, 171
Grey's Anatomy, 18
Griffin, Georgia, 102
Guy, Jasmine, 122
Gwinnett County Detention Center, Lawrenceville, Georgia, 12, *60*, 62, 95
Gwinnett County Police Training Center, 59

Hall, Stuart, 122
Hamilton, Lisa Gay, 64
Hansen, Miriam, 176
Hardison, Kadeem, 65, 144–45
Harjo, Laura, 48
Harjo, Sterlin, 52
Henderson, Felicia D., 21, 124–25

Immigration and Customs Enforcement (ICE), 70, *70*, *71*, 87, 164
Indian Day School, 49
In Plain Sight, 31, 51, 57, 59, 62, 68
Intrenchment Creek Trailhead, 37, 47–48
Isleta Pueblo, 49
Isleta Resort and Casino, 49

Jacobson, Brian R., 46
jail sets, 57
Journal Studio Center, Albuquerque, 44

Keaton, Buster, 46–47
Kemp, Biran, 96
Killer Women, 57, 62–63, 65
Ku Klux Klan, 9, 99–100, 102–4
Kwok, Miranda, 20, 69–70

labor: crime scene cleaners, 68–72; low wage, 16, 18–19, 28, 42–46, 59, 89, 104, 127, 139, 154, 167–68, 176, 178; set cleaning workers, 6
labor exploitation, 6–7, 9–10, 42–46, 88–89, 93, 154, 157–58, 189
labor unions, 7, 71, 131, 153, 165–66, 177
Lake Lanier, Georgia, 77–78
Las Vegas, New Mexico, 84–85, 87, 97, 108–9
Latinx creatives, 18–21, 23, 44, 69
Latinx racial capitalism in place, 19, 22–24, 69, 113, 119, 167
Lawrenceville, Georgia, 94, 101; Gwinnett County Detention Center, 12, 59–60, *60*, 62, 95
Lawrenceville Jail, 59

Lee, Robert E., 98, 100, 170
Lemus, Marvin, 167
Lewis, Jon, 122
Lipsitz, George, 32, 114
Longmire, 56, 84, 108
Longoria, Eva, 18–19, 44
Los Angeles Dodgers, 171
Los Angeles Police Department, 63, 66–67, 153
Los Angeles Times, 6–7, 160
The Lost Room, 32, 38–40, *58*
Lovecraft Country, 48, 56, 89–92, *90*, 125
*Love Is_*, 140–41

MacGruber, 58, *58*
Macon, Georgia, 102, 202
Madea franchise, 126
MAGA movement, 21, 55, 64–65, 73–75, 97, 123, 164, 174–75
A Man in Full, 61
Mariachi Plaza, Boyle Heights, Los Angeles, 153–54, 159, 161
Marvel Studios, 33, 52, 91, 95, 148–49
Mayans MC, 153
Mayer, Vicki, 3–5, 7, 12–13, 47, 146
Melamed, Jodi, 10
melodramas, 26, 82, 121–26, 141–42
Mesa Del Sol Housing Development, 37, 49, 147, *147*, 151, 168
The Messengers, 31–32, *32*
methamphetamine, 31, 43, 45, 49, 50, 56, 75
Midnight, Texas, 32, 38, 40, 49, 83, 83–85, 87, 93, 108–9
migrant detentions centers, 10–111, 113, 164
Moraga, Cherríe, 167
Muscogee (Creek) Nation, 25, 47–48, 51–52, 168
mylar blankets, 70

Nabers, Janine, 20, 78
Naomi, 98, 105–6, *107*

Narcos on TV, 1–2, 18–19, 25, 29, 30, 55, 65–66, 68–69
National Treasure: The Edge of History, 88
Navajo Nation, 48–49
neoliberalism, 126, 141
Netflix Albuquerque Studios, 11, 33, 36–37, 42, 49, 51, 56, 59, 84, 147, 151, 168
New Mexico Correction Department Central Administration, 56
Newnan, Georgia, 31, 102
New Orleans, 11–13
Newsome, Bree, 97–98
Nicotero, Greg, 149
Nixon, Richard, 103

Obama, Barack, 82, 135
Obama, Malia, 82
Obama, Michelle, 141
Obliterated, 147
O'Dowd, Chris, 28
Oklahoma!, 48
Old Atlanta Prison Farm, 56
Old Main Penitentiary, Santa Fe, New Mexico, 57–58, *58*, 60, 168
Opportunity: A Journal of Negro Life, 9
Oprah Winfrey Network (OWN), 118–23, 129, 141–43
Ornales, Sierra Teller, 169
Oscarville, Georgia, 77–78
Outer Range, 113–17, *116*, *117*
The Outsider, 60
The Oval, 134–35
Ovarian Psycos, 153, 161
OWN Network. *See* Oprah Winfrey Network
Ozark, 30, 34, 60, *60*, 78, 127

Penitentiary of New Mexico, Santa Fe, 59–60, 112–13
Perpetual Grace, LTD, 58
Perry, Tyler, 37, 57, 126–35, 149, 150–51, 179
Podemski, Tamara, 114–17

police and prison televisual complex, 55, 69–70, 88, 93, 97, 119, 132–33, 151
police TV advisors, 53–54, 56
Preacher, 41, 85–87, *86*, 93, 95
private planes, as status symbols, 128–29
professional sports franchises, 14–15, 132
Pueblo-style architecture, 17, 36, 75

The Quad, 20–21, 26, 119, 122–25
Queen of Outer Space, 112

racial capitalism in place, 2–3, 8–15, 44, 76, 104, 180; Black, 22–24, 80–82, 91, 119, 121–22, 126; as carceral, 25, 40–41, 55–56, 67, 71, 73; Latinx, 19, 22–24, 69, 113, 119, 167; Mexican narcos as, 18, 25, 55; violence and, 86, 89, 93, 96, 143
railroad industries, 9–10, 38–41, 63, 109, 155–60; as location warehouses, 31–36, 84–85
Raising Dion, 105–6
Rectify, 60
Redus, Norman, 149
Reservation Dogs, 51–52, 114
respectability politics, 137–38, 141–43, 164
Rimes, Shonda, 18–19
Robinson, Cedric J., 8–10, 22–24, 82, 119, 122, 157
Rodríguez, Dylan, 66–67
The Rookie, 153
Roswell, New Mexico, 109–13
Rutherford Falls, 169, 169–71
Ruthless, 127

Saigon, 51, 102, 103–4
Sangre de Cristo Mountains, New Mexio, 49
Santa Fe, New Mexico, 26, 56, 73, 75, 88, 97; Old Main Penitentiary, 57–58, *58*, 60, 168; Penitentiary of New Mexico, Santa Fe, 59–60, *60*, 112–13
Santa Fe Studios, 36–37, 56, 112–13
Saracho, Tanya, 20, 160–61, 167

Schmieding, Jana, 169–71
Senoia, Georgia Police Department, 61, 151
serial killers, 65, 73–82
set cleaning workers, 6
settler colonialism, 15, 28, 46–52, 75–76, 78, 97, 103, 113, 117, 145, 167–72
Shadowbox Studios (formerly Black Hall Studios), 5, 11, 36, 36–39, 39, 47–48, 56–57, 59, 92, 156
Sistas, 123, 127
Sleepy Hollow, 87–89, 92
Snellville City Jail, 59
Sohn, Sonia, 64
South River Watershed Alliance, 37, 57
Spigel, Lynn, 35
The Staircase, 60
Stranger Things, 35, 56
Succession, 56
Sundowner Motel, Albuquerque, 85–86, 86
supernatural DEI horror, 82–92
Swarm, 78–81, 81

teachers, 40, 107–8, 122, 128
Teenage Bounty Hunters, 65, 98, 140
Terminator: The Sarah Connor Chronicles, 32
"territorial revival style" of architecture, 17, 74–75
Texas Rangers, 64–65
textual methods of interpretation, 13–14, 16, 105
To'hajiilee, New Mexico, 49–51, 50
Tongva Nation, 169, 171–72
trans characters, 63, 66, 111

Treme, 11–13
Trilith Studios, Fayetteville, Georgia, 147–48, 148
Trump, Donald, 19, 24, 66, 75, 88, 99, 174
TV shows about TV shows, 16–17, 123, 135–41, 156–58, 164–65, 177

Union, Gabriel, 122, 136
United Farm Workers, 165–66
United Talent Agency, 161
Unite the Right rally, 107, 123, 170
Universal Studios, 171
The Unsettling, 73–75
Ursula Detention Center, McAllen, Texas, 70

Vida, 160–63
Vietnam War, 103–4

Walker, Alice, 124
Walker: Independence, 63–64
The Walking Dead, 31–33, 40, 48, 57, 61, 72, 92, 145, 149–51, 150, 154–56
"Wash Her Out," 162
Watchmen, 79, 98, 98–105
West of Zanzibar, 9
Will Trent, 11–12, 12, 60, 64–66
Winfrey, Oprah. *See* Oprah Winfrey Network
Womack, Craig, 47
Wood, Robin, 40, 93–94
Woods, Clyde, 72
Writers Guild of America, 177

Yung, Elodie, 44

ABOUT THE AUTHOR

A Professor in the Ethnic Studies Department at UC San Diego, CURTIS MAREZ is the author of *Drug Wars: The Political Economy of Narcotics*, *Farm Worker Futurism: Speculative Technologies of Resistance*, and *University Babylon: Film and Race Politics on Campus*. He is the former Editor of *American Quarterly* and former President of the American Studies Association.

www.ingramcontent.com/pod-product-compliance
Lightning Source LLC
Chambersburg PA
CBHW031146020426
42333CB00013B/536